# THE ECONOMICS OF THE LATECOMERS

Is 'backwardness' a disadvantage? Alexander Gerschenkron developed a model of economic development in which less-developed countries are not necessarily at a disadvantage to more advanced countries. The application of new strategies and the emergence of new institutions could systematically compensate for inadequate supplies of capital, skilled labour, entrepreneurship and technology found in the more advanced economies. With this in mind, *The Economics of the Latecomers* attempts to explicate the 'catching-up' process of the most spectacularly successful economies of East Asia, Japan and South Korea.

Combining insights from economic history, development economics and the economics of technology, the book emphasises the need for historical models to understand historical processes. This perspective enables the author to demonstrate the limitations of neo-Schumpeterian approaches and the New Institutional Economics as means of analysing the development process.

The comparison of the 'catching-up' process in Japan and South Korea includes studies of the iron and steel and semiconductor industries. These demonstrate the difficulties involved in trying to detect a general pattern, as both industries appear to respond to different technological imperatives. As a result, the author argues that general models of development should be treated with caution, and highlights the importance of learning by doing and of the differing institutional contexts.

**Jang-Sup Shin** is currently working as a journalist for the Maeil Business Newspaper and TV in Seoul. He received his M.Phil. and Ph.D. from Cambridge University.

First published 1996
by Routledge
11 New Fetter Lane, London EC4P 4EE

Simultaneously published in the USA and Canada
by Routledge
29 West 35th Street, New York, NY 10001



Typeset in Garamond by LaserScript, Mitcham, Surrey
Printed and bound in Great Britain by
Biddles Ltd, Guildford and King's Lynn



*British Library Cataloguing in Publication Data*
A catalogue record for this book is available from the British Library

*Library of Congress Cataloguing in Publication Data*
A catalogue record for this book has been requested

ISBN 0–415–14055–2

# CONTENTS

# FOREWORD

The uneven development of the world economy is a feature which has attracted much attention from economic historians as well as political scientists and politicians. The American economist Abramovitz in particular has coined the expressions 'forging ahead', 'falling behind', and 'catching-up' to designate phases in the history of nations which call for explanation in terms of a theory of economic growth. In his designation, Britain 'forged ahead' of the rest of the world economy during the Industrial Revolution of the late eighteenth and early nineteenth centuries, while the USA forged ahead in the late nineteenth and early twentieth centuries.

Much more common but ultimately of great importance for the world economy is the phenomenon of 'catch-up' by the less developed countries. Many European countries overtook and in some cases surpassed Britain in their rate of growth and their productivity in the twentieth century, whilst Europe and Japan caught up with the USA in many respects in the decades which followed the Second World War.

This book is about 'catch-up' and throws new light on the whole process of catching-up in the world economy. The author follows Gerschenkron in arguing that latecomers actually have some advantages in the catching-up process, deriving from the very fact of their backwardness. Like Gerschenkron, Jang-Sup Shin takes the steel industry as an exceptionally clear illustration of the catching-up process. Latecomers were able to import and exploit the technologies already developed elsewhere and, in addition, to derive extra scale economies from leapfrogging in plant size. The author illustrates these points from the experience of the German steel industry, but of even greater interest is his analysis of catching-up by Japan and Korea.

His analysis of the Korean catch-up is an extremely valuable contribution to our understanding of world economic development in the most recent decades. Using the examples of the semiconductor industry and the steel industry he shows the great importance of various social *and* organisational innovations in Korean catching-up. Latecomers have a *potential* for rapid growth in overtaking the leaders, but the *realisation* of this potential depends upon a process of institutional innovation facilitating the process.

audience would be general readers, not merely academics. It would not be necessary for my work to be geared towards a general audience from the beginning. It should, however, be at least an intellectual synthesis which could communicate to a broader audience by connecting academic research with journalism. And my target audience would be mainly Koreans, since my paper is clearly based on national boundaries.

The above was a prime consideration when I chose development economics as my initial field of research. I expected that, by studying development economics, I might be able to view the economic development process of South Korea in perspective and thereby come to see some implications for its future growth. In my first year at Cambridge, which was a taught course, I learned a good deal from my teachers and friends. I was, however, disappointed by the fact that development economics was treated as peripheral to current economics. My impression was that mainstream economics claimed to deal with general aspects of economic phenomena which are supposed to be more prevalent in advanced countries than in developing countries while development economics was supposed to deal with more specialised aspects of economic phenomena. In a sense, development economics was no more than a special application of *economics*.

However, my journalistic experience told me that economic processes are essentially dynamic and include complex forces. I thought that the experiences of developing countries had no less to tell us about these dynamic processes than those of advanced countries. Moreover, advanced countries had been developing countries at some point of their history. I wholeheartedly agreed with Albert Hirschman (1986: 14) when he confesses his discovery of general applicability of development economics.

> It appears . . . that the very characteristics on which I had sought to build an economics specially attuned to the underdeveloped countries have a far wider, perhaps even a universal, range and they define, not a special strategy of development for a well-defined group of countries, but a much more generally valid approach to the understanding of change and growth. In other words, I set out to learn about others, and in the end learned about ourselves.

In retrospect, what I have attempted in this book is to put the development experience of South Korea into a general perspective. I was fortunate to encounter Alexander Gerschenkron in the course of my research. Born in Russia, he was mainly concerned with interpreting Russian industrialisation which had begun in earnest from the late nineteenth century. I was mesmerised by his intellectual attempt at putting his native country's experience into an encompassing perspective of the European industrialisation process in the nineteenth century.

This book is basically an elaboration and extension of Gerschenkron's insight into the twentieth century. I have attempted to apply his schema to interpreting the development process of Japan and South Korea during the post-war period. Although I have also attempted to interpret Prussian industrialisation in the first half of the nineteenth century with Gerschenkron's schema, the main thrust of my analysis is on explaining the experiences of Japan and especially South Korea.

In suggesting Gerschenkron's schema as an interpretative tool for explaining late industrialisation, I feel obliged to clarify some methodological points on which this book was premised. The first three chapters of this book are devoted to methodological discussions. I have also narrowed my focus to the technological side of late industrialisation because technological imperatives and subsequent processes of technology transfer seem to have much to say in explaining patterns of industrialisation. Neo-Schumpeterian studies of technology and economics are discussed in this context. With the historical interpretations advanced here, I have tried to address several theoretical issues raised in development economics as well.

As a result, this book has become a combination of methodology, economic history, technology economics and development economics. I hope it does not read as a hotchpotch of different ideas. I will be grateful if readers find something both holistic and realistic in the connection of these diverse fields.

In writing this book, my greatest thanks are due to Mushtaq Khan who has guided me with constant intellectual stimulation and warm encouragement from the beginning of my time at Cambridge. He was not only my supervisor but one of my closest friends. His constant emphasis on theory and history in one kept me from crude generalisation or simple factual description.

I am also deeply indebted to Ha-Joon Chang who happened to be my teacher at Cambridge although he was my junior in our BA course in Seoul. He helped me a lot and gave me invaluable advice. He has also carefully read most of my drafts and given me detailed comments. Without his incisive comments this book would have looked very different.

I wish to record my special thanks to Professor Christopher Freeman who was my external examiner twice, both for M.Phil. and Ph.D. dissertations. Since he examined my M.Phil. work which was completed in 1992, he became, in my mind, the target reader for the Ph.D. dissertation. It was a great incentive to my research that I should show him notable improvements on the M.Phil. work. He came over to Cambridge to conduct my Ph.D. exam although he was in bad health. Moreover, his enthusiastic recommendation was crucial to the transformation of my dissertation into this book.

My understanding of European economic history owes much to Clive Trebilcock's stimulating teaching. His constant reference to historical models in explaining historical details was of great help to me. Young-Jin

ground, political conditions, natural resources, country size, population, and so on. A main focus of this book is on the difference in the time element following Gerschenkron's (1962; 1968; 1970) insight. In our view, the development process of latecomers is, in significant respects, different from that of forerunners simply because it is late. Forerunners develop without a reference point. They have to make their own way by themselves. But they are entitled to 'supernormal profits' due to the absence of existing competitors. By contrast, latecomers develop with constant reference to forerunners. They can learn a lot from the experience of forerunners. But they have to develop facing formidable competition from existing forerunners. The existence of forerunners has both complementary and competitive aspects to latecomers' development. If forerunners and latecomers do not interact with each other in the same system, it does not make much sense to distinguish between forerunners and latecomers. But it seems to us that this interaction mechanism is a strategically important factor in understanding the process of late industrialisation.[4]

Once one begins to view late industrialisation as above, catching-up is no longer a puzzling phenomenon. Rather, it is part and parcel of the general process of economic growth in which forerunners and latecomers interact, with their respective advantages and disadvantages. According to different combinations of (dis)advantages of latecomers and those of forerunners, forging ahead or catching-up can be the result. On balance, catching-up is no more a puzzling phenomenon than forging ahead is.

Moreover, one could even argue that the phenomenon of forging ahead has been more exceptional than that of catching-up. There have been only two cases of obvious forging ahead in modern industrial history: that by Britain from the late eighteenth century to the middle of the nineteenth century and that by the USA during the first half of the twentieth century. By contrast, the process of catching-up has been followed by a much larger number of countries, though the pace of new entrants might have been disappointingly slow for the aspirations of those in the underdeveloped countries. And, as far as its quantitative impact on the overall economic growth of the world economy is concerned, catching-up has had a much greater impact than forging ahead.[5]

Chapter 1, 'Late industrialisation and catching-up', is devoted to clarifying methodological issues concerning the catching-up debate. Broadly categorising existing discussions about catching-up into three types, an assessment of their methodological underpinnings is attempted. Abramovitz's attempt at analysing the catching-up phenomena, which relies on such broad concepts as social capability, potential, realisation, and so on, will be critically assessed. Perez and Soete's discussion of technological entry barriers will also be examined critically. And we shall highlight the significance of Gerschenkron's view of catching-up.

The lateness in time is reflected in several things like stock of wealth, technology, institutions, standard of living, educational level, and so on. It is practically impossible to deal with all of them in this short monograph, despite the fact that they are closely interrelated. The specific focus of this book is technology. One main impetus behind the forging ahead by the UK and the USA was their technological advancement. By the same token, the rapid catching-up by other latecomers was greatly helped by the utilisation of existing technologies mostly created by their forerunners. Chapter 2, 'Technology transfer, technological capability and national systems of innovation', is concerned with questions such as what is the mechanism of technology transfer, what are strategic factors affecting it, how can the process of technology transfer be analysed, and so on.

The neo-Schumpeterian economists in particular have contributed significantly to deepening our understanding of the process of technology transfer. It seems, however, that their contribution is still largely confined to empirical studies, and some analytical concepts proposed by them are not quite securely established. Particular attention is paid to their concepts of 'technological capability' and 'national systems of innovation'. We argue that these concepts in themselves are still too broad to help increase our understanding of the process of technological progress in latecomers. We need to develop lower-level analytical tools which guide us in explicating the interaction among the components constituting these concepts if we want to reach a meaningful causal imputation.

Chapter 3, 'Institutions, institutional change and catching-up', is devoted to developing an analytical framework for our empirical study of the catching-up process. At first, we emphasise the inevitable tension between theory and reality. And Max Weber's methodology of constructing an 'ideal type' and its application to understanding the actual historical process is highlighted as a way to accommodate this tension. In discussing the relevance of 'transaction cost economics' in comparing economic performance across countries, it will also be emphasised that a schema relying only on universal concepts has difficulty in overcoming the tension between theory and reality.

The focal point of this book is the relation between technology and institutions in the process of catching-up. We find it more profitable to pay attention to the differences of institutional responses than to their similarities across countries. This is mainly because institutions function in their context, which is quite diverse across countries. We pay special attention to Gerschenkron's 'pattern of industrialisation' as an analytical framework accommodating the diversity across countries in the process of catching-up. His schema is directed at explaining different patterns of industrialisation among different latecomers with 'the degree of backwardness' as an organising concept.

Gerschenkron's schema was, however, based on his observation of

# Part I

# ON METHODOLOGY

*Table 1.1* Convergence and divergence among the OECD countries: changes in comparative levels of per capita GDP
(US per capita GDP = 100)

| *Country* | *1820* | *1870* | *1913* | *1950* | *1973* | *1989* |
|---|---|---|---|---|---|---|
| Australia | 119 (88) | 138 (119) | 93 | 69 | 73 | 74 |
| Austria | 99 (74) | 64 (55) | 55 | 33 | 61 | 69 |
| Belgium | 98 (73) | 93 (80) | 66 | 49 | 67 | 70 |
| Canada | n.a. | 60 (52) | 73 | 71 | 84 | 96 |
| Denmark | 94 (70) | 69 (60) | 63 | 61 | 75 | 74 |
| Finland | 61 (45) | 42 (36) | 36 | 40 | 64 | 76 |
| France | 100 (75) | 70 (60) | 56 | 48 | 73 | 76 |
| Germany | 89 (67) | 58 (50) | 54 | 39 | 72 | 76 |
| Italy | 92 (68) | 54 (46) | 42 | 33 | 61 | 71 |
| Japan | 56 (42) | 28 (24) | 23 | 18 | 65 | 82 |
| Netherlands | 125 (93) | 92 (79) | 65 | 55 | 73 | 70 |
| Norway | 82 (61) | 53 (46) | 43 | 53 | 66 | 90 |
| Sweden | 90 (67) | 59 (50) | 50 | 62 | 80 | 81 |
| Switzerland | n.a. | 82 (71) | 63 | 76 | 93 | 84 |
| UK | 134 (100) | 116 (100) | 83 | 66 | 71 | 74 |
| USA | 100 (75) | 100 (86) | 100 | 100 | 100 | 100 |
| Average | 95 (69) | 72 (63) | 58 | 52 | 72 | 78 |

*Source*: calculated from Maddison (1991, table 1.1)

*Notes*: 1 numbers in parentheses represent comparative per capita GDPs when that of UK is 100
2 average means the arithmetic average of fifteen countries excluding the base country

This convergence process, however, was not extended outside the OECD countries. Low- and middle- income countries, commonly called 'developing countries', in general did not catch-up with high-income countries. As Table 1.3 shows, the average annual growth rate of per capita GNP of low- and middle-income countries during 1965–1989 was 2.5 per cent, which was the same as that of the OECD countries. Consequently, the income gap between country groups did not shrink. In 1989, per capita GNP of low- and middle-income countries ($800) remained less than 5 per cent of the OECD countries' ($19,090).

There is also regional and temporal disparity within developing countries. It may appear from Table 1.3. that low-income economies with an average growth rate of 2.9 per cent have been narrowing the gap with high-income economies. But this is mainly due to the unusually high growth rate of China, which contains around 38 per cent of the whole population of

*Table 1.2* Convergence and divergence among high-income countries: changes in comparative levels of productivity
(US GDP per man-hour = 100)

| *Country* | *1870* | *1913* | *1950* | *1973* | *1987* |
|---|---|---|---|---|---|
| Australia | 132 | 93 | 67 | 70 | 78 |
| Austria | 51 | 48 | 27 | 59 | 74 |
| Belgium | 82 | 61 | 42 | 64 | 86 |
| Canada | 64 | 75 | 75 | 83 | 92 |
| Denmark | 59 | 58 | 43 | 63 | 68 |
| Finland | 34 | 33 | 31 | 57 | 67 |
| France | 56 | 48 | 40 | 70 | 94 |
| Germany | 50 | 50 | 30 | 64 | 80 |
| Italy | 41 | 37 | 31 | 64 | 79 |
| Japan | 19 | 18 | 15 | 46 | 61 |
| Netherlands | 88 | 69 | 46 | 77 | 92 |
| Norway | 48 | 43 | 43 | 64 | 90 |
| Sweden | 47 | 44 | 49 | 76 | 82 |
| Switzerland | 62 | 51 | 56 | 67 | 68 |
| UK | 104 | 78 | 57 | 67 | 80 |
| USA | 100 | 100 | 100 | 100 | 100 |
| Average[1] | 62 | 54 | 43 | 66 | 79 |

*Source*: Maddison (1991, table 3.4)

*Note*: 1 arithmetic average of fifteen countries excluding the USA

low-income countries.[2] Its per capita income has grown at a rate of 5.7 per cent per annum during the period. If China is excluded, the average growth rate of other low-income countries is less than 1.8 per cent, far lower than that of the OECD countries.[3] The poorest countries, mostly situated in Sub-Saharan Africa and South Asia have been falling behind. Per capita GNP of these two regions grew at 0.3 per cent and 1.8 per cent respectively. The only region which has narrowed the income gap with high-income countries is East Asia. The average growth rate of per capita GNP in this region has been more than twice that of the OECD countries. Latin American and Caribbean countries seemed to have narrowed the income gap with the OECD countries before 1980, but they began rapidly falling behind afterwards.

As far as the factual descriptions of 'catching-up, forging ahead and falling behind' above are concerned, there is little disagreement among historians or economists.[4] The trouble is with its interpretation. Various attempts have been made to explain the implications of the picture described above. Yet, it does not seem that the methodological under-

country towards the beginning of the First World War. The second version of the catch-up hypothesis indicates that convergence should have occurred from that time. But it did not happen because of 'the irregular effects of the Great War and of the years of disturbed political and financial conditions that followed, by the uneven impacts of the Great Depression itself and of the restrictions on international trade'.[9] Rather, during the interim period of 1913–1950, a rapid growth potential in the follower countries was further accumulated as the USA forged ahead. Convergence became strong only after the Second World War, with the emergence of the international conditions favourable to catching-up.

Hence, Abramovitz's third qualified version of the catch-up hypothesis: 'The pace at which potential for catch-up is actually realised in a particular period depends on [realisation] factors limiting the diffusion of knowledge, the rate of structural change, the accumulation of capital, and the expansion of demand.'[10] For Abramovitz, the match between the accumulated (total) potential before the end of the Second World War and the emergence of realisation factors in the post-war period explains 'the Golden Age of Capitalism'.

From these qualified versions of the catch-up hypothesis,[11] interpretations of the failure cases follow. As regards the reasons for the absence of convergence among the OECD countries during the nineteenth century, Abramovitz explicitly points to the divergence in social capability: 'social competence for exploiting the then most advanced methods was still limited, particularly in the earlier years and in the more recent latecomers.'[12] He does not explicitly discuss the applicability of his view to the current developing countries. But he implicitly suggests that the low level of social capability in these countries hindered them from exploiting the existing technology gap, although realisation factors were favourable during the post-war period.[13]

Other studies of the catch-up hypothesis stand on the same position. For instance, Baumol (1986) suggests that low levels of industrialisation and education in poorer less developed countries have barred them from utilising the technology gap. Dowrick and Gemmel (1991) confirm Baumol's suggestion by using a regression model.

A characteristic feature of Abramovitz's analysis is the employment of the averaging method in detecting principal causes.[14] To rely on the average is useful, especially when one deals with a large sample. But this method has its own limitations, one of which is the tendency that the interpretations attained from the cases which have a significant impact on the average are often applied to other cases which are not detected by the average. This practice assumes, implicitly or explicitly, a universal set of cause and effect across countries.

In Abramovitz's analysis, what actually has a significant effect on the aggregate picture of his sample is the post-war catching-up process. During this period, it is plausible to regard the high level of social capability as a cause for catching-up *common to all the OECD follower countries*. But this interpretation relies crucially on *the accidental situation* that the level of social capability was separated from that of technology, that is, the situation that a large technology gap exists without a significant gap in social capability among the OECD follower countries.

A problem arises, therefore, if one tries to apply this interpretation to other periods or other countries. As we pointed out above, Abramovitz explains the absence of catching-up in the nineteenth century among the OECD countries by the disparities in social capability. But how meaningful is this interpretation? Does it make sense to say that the latecomers in the nineteenth century could not narrow the technology gap with their forerunners because they lacked sufficient social capability?

There were also local cases of technological catching-up which did not appear in the aggregate picture. Although data of per capita income or labour productivity do not show convergence among European countries in the nineteenth century (Tables 1.1 and 1.2), it is widely documented and accepted that some European countries, such as Germany, France and even Russia, rapidly narrowed their technology gap with Britain. Germany even overtook Britain in certain industries, including steel and chemicals. Likewise, the East Asian NICs have rapidly narrowed their technology gap with advanced countries in spite of the general failure of catching-up by the developing countries during the post-war period. What can be said about these local cases of catching-up? Is it plausible to say that they were successful in catching-up because they had already enlarged their social capability successfully *before* they began reducing the technology gap with their forerunners?

These are essentially questions of causal imputation.[15] If one tries to apply the concept of social capability in explaining the cases of catching-up failures or those of local catching-up as above, one in effect assumes a staged development leading from the acquisition of social capability to technological catching-up. Here, the process of enlarging social capability is separated from that of technological catching-up, and the former is regarded as a general precondition for the latter across time and space.

One might argue that this issue can be satisfactorily resolved by further empirical research. But this is unlikely for the following reasons. First of all, the concept of social capability is ambiguous. Its meaning may be 'sufficiently clear' only during the post-war period among the OECD countries, as Abramovitz argues.[16] In general, however, it does not seem to give us a definite enough meaning to resolve empirical questions. Since we shall discuss the conceptual ambiguity of social capability in detail later (section 2.1), we just want to point out here that Abramovitz (1986: 222) himself admits that '[t]he trouble with absorbing social capability into the catch-up hypothesis is that no one knows just what it means or how to

being a latecomer. But their works have been no more than descriptive presentations of the diverse elements of advantages and disadvantages without assessing the 'net' sum of the late-comer's (dis)advantages at a particular period or for a particular group of countries.[25] An ingenious aspect of Perez and Soete's analysis is their attempt to solve the conundrum of the 'net advantages' of latecomers by employing the concept of 'techno-economic paradigm'. In short, their contention is that, if technological and economic development is viewed in terms of the techno-economic paradigm, it is a cumulative and continuous process *within* a paradigm but a destructive and discontinuous process *across* paradigms. During the period when the dominant techno-economic paradigm is stable, technological and economic development takes the characteristic of increasing returns because technologies are clustered within a paradigm and because institutions have evolved to support the paradigm. So, an early advancement facilitates further advancement. The technological and institutional advantages of forerunners tend to be sustained and strengthened within a techno-economic paradigm.

Of course, Perez and Soete acknowledge that there are product life-cycles, as proposed by Hirsch (1965) and Vernon (1966) within a techno-economic paradigm, and that these affect the latecomer's catching-up opportunities. Technologies of a product become standardised as they mature, and accordingly comparative advantage shifts in favour of less developed countries. This opens a window of opportunity for latecomers to catch-up with forerunners by entering the production of standardised products. Let us call it 'the first window of opportunity'. But Perez and Soete do not grant much importance to this first window. They argue that the technological and institutional dynamism of forerunners is not damaged by transferring some mature products to less developed countries. That is, they contend that forerunners retain net aggregate advantage over latecomers within a techno-economic paradigm even with some technology transfer to latecomers.

For Perez and Soete, the real window of opportunity for latecomers opens during the period of techno-economic paradigm shifts. Previous technological and institutional development no longer acts as an advantage in this transition period. Rather, it becomes a costly burden to the structural adjustment of forerunners. On the other hand, latecomers are in an advantageous position, since they do not possess the burden of capital stocks and institutions inherited from the past. They are 'lighter and faster' in a new technology system. This advantage of latecomers is corroborated by the possibility that public knowledge is available in the early phase of a technological system. Therefore, latecomers can gain net advantage over forerunners during the period of techno-economic paradigm change. Let us call it 'the second window of opportunity'.

This kind of grand generalisation is very rare in neo-Schumpeterian studies. Most of them are more concerned with specific characteristics of technologies and their implications for development strategy. And indeed, criticisms of Perez and Soete from other neo-Schumpeterians have come from their studies of specific technologies. For instance, Ernst and O'Connor (1989) question whether the windows of opportunity are really widely open in the transition period of the current microelectronics (ME) revolution. They argue that the ME revolution brings about increasing flexibility in production, increasing economies of scale, increasing science intensity, and so on, and accordingly makes it more difficult for developing countries to catch-up with advanced countries.[26] Kaplinsky (1989) also contends that the technology gap between advanced and developing countries is more likely to be reopening with the transition of production systems from 'machinofacture' to 'systemofacture'.

We can add two other criticisms to Perez and Soete's work. First, their analysis does not seem to quite fit the actual history of catching-up. Perez and Soete seemingly deal with the long-term historical process by referring to the industrialisation of European countries and the USA in the nineteenth century, and of Japan and South Korea in the twentieth century. But a detailed analysis of those historical experiences is absent. Rather, it is implicitly argued that those successful catching-up cases were principally brought about during the periods of techno-economic paradigm shifts.[27]

Second, and more fundamentally, catching-up is an historical phenomenon emerging *in spite of* technological entry barriers, not a result of some inherent characteristics of technologies or technological trends. This would be an important reason why predictions based solely on some supposed objective characteristics of technologies, as can be seen in dependency theory, do not fit historical reality. It is certainly important to analyse the characteristics of technology gaps and technological trends – without such analysis any scientific formulation of a catching-up strategy is impossible. But the utility of that kind of analysis lies primarily in indicating scope and direction of responses on the part of latecomers. The actual catching-up is determined by the combination of the characteristics of technologies and responses on the part of latecomers. If one goes too far with the concept of technological entry barriers, one is liable to fall into technological determinism.

Most neo-Schumpeterian economists are cautious when they draw implications from their analyses of specific technologies. They often stress the existence of niche markets for latecomers even when the entry barrier looks extremely high. Yet they do not seem entirely successful in detaching themselves from technological determinism.[28] Perez and Soete's neglect of the first window of opportunity is a case in point. They explicitly argue that 'in so far as mature products are precisely those that have exhausted their technological dynamism, this choice implies clear risk of getting "fixed" in low wage, low growth, development pattern.'[29] This contention is derived from their emphasis on the system characteristic of the techno-economic paradigm that 'the knowledge, the skills, the experience and the externalities required for the various products within a system are interrelated and support each other'.[30] In this *a priori* description of the system, there is no room for latecomers to catch-up with forerunners. Latecomers are unavoidably disadvantaged in this system.[31]

It also seems important to point out that there is an implicit assumption about the relation between technological and institutional change in Perez and Soete's analysis. In their concept of the techno-economic paradigm, where compatibility between the technological and institutional system is emphasised, the latter is largely treated as a dependent variable. It is certainly necessary that institutions should continuously adapt themselves to constantly changing technological environment, which can be regarded as an autonomous force from the viewpoint of individual latecomers, but sources of institutional change in a country are diverse and not confined only to external technological change. For instance, institutional change can occur in order to exploit existing technologies or anticipated technological changes, or to overcome constraints imposed by technological trends. An important mistake of Perez and Soete is their attempt to treat technological change and its connections with the economy as a *universal cause* for technological catching-up across countries. By jumping from an analysis of technological trend to the general prospect of latecomers' catching-up, Perez and Soete's analysis fails to avoid technological determinism.[32]

In this respect, it seems important to note that Rosenberg (1982: 270), in one of the few studies of the long-term historical process of technology transfer, does not give much importance to the question of entry barriers, as seen in the following quotation:

> One central conclusion that may be drawn from this account is that the transfer of industrial technology to less developed countries is inevitable. Indeed, as we have seen, the process has already been going on for about a century and a half, and there is no compelling reason to believe that it will stop . . . Thus, a historical perspective suggests that the central questions are not whether industrial technologies will be transferred, but rather when it will happen, where it will happen, which technologies will be transferred, how they will be modified in the process, and how rapidly this process will occur. One of the most interesting aspects of the history of technology transfer is the difficulty of controlling or containing its speed.

The phenomenon of catching-up can be properly understood only by combining the analysis of technological entry barriers with that of responses in latecomers.

## 1.3 GERSCHENKRON: A DIFFERENTIATED SYSTEM

Gerschenkron's approach to catching-up stands in stark contrast to Abramovitz's. Although Abramovitz mentions the possibility of the interaction between the technology gap and social capability, the processes of enlarging social capability and that of catching-up are separated in his main analysis. Due to this separation, his explanation becomes rather simple: catching-up is determined by social capability subject to realisation factors. In contrast, Gerschenkron starts his analysis from the coexistence of the technology gap and the social capability gap, and therefore emphasises the simultaneity of enlarging social capability and technological catching-up. Here, the explanation is not so simple.

First of all, Gerschenkron stresses that the crucial features of the late-comer's development are different from those of the forerunner's simply because they are late. His criticism of Rostow's view of stages of economic growth or the Marxist conception of 'original accumulation of capital' is concerned with this issue.[33] Of course, he acknowledges that 'there are common features in all industrializations' and that '[t]o concentrate upon these general aspects of industrialization may be quite useful for some purposes'.[34] But Gerschenkron points out that they explain, at best, 'the half-truth', and that the emphasis on general factors quite often results in a 'curious' view of 'prerequisites' for economic development:

> Certain major obstacles to industrialization *must* be removed and certain things propitious to it *must* be created before industrialization can begin. Both in its negative and its positive aspects, the concept seems to imply . . . at least the notion that it must proceed in a certain manner; that is to say, through certain more or less discrete stages. Along with it goes the idea of the uniformity of industrial development in the sense that every industrialization necessarily must be based on the same set of preconditions.[35]

However, Gerschenkron's observation of European industrialisation in the nineteenth century does not fit this view. He points out the following.

> (1) Some of the factors that had served as prerequisites in the advanced country either were not present at all, or at best were present to a very small extent, in the more backward countries. (2) The great spurt of industrial development occurred despite the lack of these 'prerequisites'.[36]

According to the 'prerequisite' view, these cases are treated simply as exceptions. But, to Gerschenkron, they are too important to disregard. Thus he opts 'to systematize the deviations from the original pattern by bringing them into a new, although necessarily more complicated, pattern'.[37]

A crucial factor in Gerschenkron's systematisation is the economic backwardness of latecomers itself. In a forerunner country like Britain, it has taken a long time for the prerequisites of industrialisation to emerge. For latecomers, however, the very existence of a more advanced country provides them with what amounts to prerequisites for the development of the forerunner. Without necessarily creating those prerequisites indigenously, latecomers can initiate their industrialisation mainly through utilising technological and institutional backlogs already created by forerunners. So Gerschenkron (1962: 46–7) argues:

> One thing is obvious. Illiteracy and low standards of education, and the resulting difficulty in training skilled labor and efficient engineers, can be overcome to some extent by immigration from more advanced countries and to some extent by using the training facilities of those countries. The same is true, even more importantly, of the lack of store of technical knowledge. It can be imported from abroad. In this sense, however, one can say that in a backward country there exists a 'prerequisite' to industrial development which 'the' advanced country did not have at its disposal, that is, the existence of the more advanced countries as sources of technical assistance, skilled labor, and capital goods.

Here, the substitution of missing prerequisites, or the utilisation of existing prerequisites, and catching-up is more of a simultaneous than a sequential process in latecomers. Gerschenkron therefore argues in several places that those prerequisites are created 'not before, but *in the very course* – and as a consequence – of a spurt of industrialization'.[38]

However, this process is not automatic. Hence his emphasis on institutions:

> industrialization processes [of latecomers] . . . showed considerable differences, as compared with more advanced countries, not only with regard to the speed of the development . . . but also with regard to the productive and organizational structure of industry which emerged from those processes. Furthermore, these differences in the speed and character of industrial development were to a considerable extent the result of application of institutional instruments for which there was little or no counterpart in an established industrial country.[39]

The central focus of Gerschenkron's analysis of late industrialisation is therefore on the question of how those often 'missing' prerequisites in latecomers are created or substituted for through specific institutional responses, rather than on whether some prerequisites exist sufficiently in particular latecomers at the beginning of their development.

He proposes to understand this substitution process coherently with 'the degree of backwardness' as an organising concept. Since we shall elaborate on his schema in detail later (section 3.2), a brief comment should suffice here. Gerschenkron's comparative schema is a tripartite system: one forerunner and two latecomers who are situated in different degrees of backwardness, i.e., that of 'moderate backwardness' and that of 'extreme backwardness', interact in a system. This schema is mainly abstracted from the historical experience of catching-up with Britain by Germany and Russia in the latter half of the nineteenth century. Here, the forerunner, i.e., Britain, sets a reference point and the latecomers, i.e., Germany and Russia, follow. The latecomers apply special institutional devices in their efforts at reducing the gaps with the forerunner *both* in technology and social capability within a short time.

In the emergence of these institutional responses in the latecomers, technological trends also play an important role in Gerschenkron's schema. He observes that technological trends in the latter half of the nineteenth century were moving towards increasing capital intensity. In the Perez-and-Soete-type analysis of technological entry barriers, this technological trend is likely to be perceived as disadvantageous to latecomers, who are usually characterised by their lack of capital. Gerschenkron, however, pays attention to the fact that the latecomers like Germany and Russia gained ground on Britain *despite* this technological trend. He argues that the latecomers in the nineteenth century could catch-up with their forerunner by undertaking a more capital-intensive investment in *individual* plants, although the overall capital intensity in their economy might be lower than that of the forerunner.

In Gerschenkron's schema, the main role of institutions in the latecomers is to mobilise resources and to concentrate them on these capital-intensive projects. Gerschenkron argues that, in a country of moderate backwardness, like Germany at the time, the investment banks assumed this role. In a country of extreme backwardness, like Russia, however, the activities of the state substituted for the role of the banks in Germany. For it was impossible in Russia to build up an effective long-term credit system because 'the standard of commercial honesty had been so low and where economic, and particularly mercantile, activities and deceit were regarded as inseparably connected'.[40]

According to Gerschenkron, the difference in the levels of social capability between the two latecomers only explains the difference in patterns of catching-up between them, not their success or failure in catching-up. Here, the initial level of social capability is no longer treated as a universal cause for catching-up. Also, the technological trend itself does not explain the success or failure in catching-up because latecomers can devise a catching-up strategy to overcome the constraints imposed by the technological trend. The causality of catching-up is explained by a more complicated system in Gerschenkron's schema, which has to do with the emergence of specific institutions and specific catching-up strategies, as will be elaborated later (section 3.2). But, at this point, we can call his explanatory schema a 'differentiated system' because differences between forerunners and latecomers, and those among latecomers, rather than common factors across countries, are regarded as crucial explanatory variables.

It should be noted that the applicability of Gerschenkron's schema is constrained just because it is a differentiated system. Abramovitz's schema, which is based on a universal proposition, claims to deal with the sixteen OECD countries for a period of nearly two centuries. Basically, the same system of explanation is extended even to the current developing countries by Baumol (1986) and others. Perez and Soete also propose their analysis as applicable to *any* latecomer. By contrast, Gerschenkron's schema is primarily derived from the European industrialisation in the latter half of the nineteenth century. Moreover, it is based only on the experiences of those countries which successfully 'caught up' within Europe. It therefore has limitations in application outside Europe or in different periods, or in the comparison between successful and unsuccessful catching-up.

These limitations are, to a significant extent, inevitable in accommodating differences across countries within an explanatory schema. As we shall elaborate later (section 3.2), Gerschenkron's schema has its own *temporal and spatial limits* as a differentiated system. If a sample is too large, it is difficult to devise a differentiated system because differences among the sample become increasingly difficult to systemise. And it becomes inevitable that one relies on universal concepts across the sample countries and periods if one wants a systematic explanation. However, these universal concepts tend to be too vague or to be reduced to some prerequisites, as seen before in the case of Abramovitz's schema (section 1.1).

As we shall further elaborate later (Chapter 3), any explanatory schema of historical process has its own limitations because there is an unavoidable tension between theory and reality. It is impossible to contain all the reality within a theory. We are required to abstract from the reality and thus any theory is unavoidably *partial*. And the nature and the extent of limitations of an explanatory schema depend basically on the level of abstraction, which is determined by the nature of the research project in question. Gerschenkron's schema is an intermediate framework situated between universal propositions and descriptive history. Its strength is in demonstrating different strategic factors across countries; its weakness lies in the limits of its application. On the other hand, Abramovitz and Perez and Soete's analyses take a higher level of abstraction. As a result, their strengths may lie in accommodating a larger number of countries, but these analyses have their own weaknesses, as we have already discussed.

## 1.4 CONCLUDING REMARKS

Abramovitz's concepts of 'potential' and 'realisation', with some modification, seem to provide us with a convenient way to synthesise the previous discussion of catching-up. Since convergence is not yet a ubiquitous phenomenon, the distinction between potential and realisation is required if one wants to establish the notion of catching-up as a general dynamic in latecomers. But what constitutes potential and realisation needs to be changed from Abramovitz's original formulation if we want to understand the process of catching-up from the viewpoint of individual countries.

Let us start with realisation, reversing Abramovitz's presentation. What is important is that his realisation factors are those affecting the *average* pace of all the latecomers' catching-up, rather than that of an individual latecomer. As we pointed out before (section 1.1), they are actually related to the factors which brought about 'the Golden Age of Capitalism' during the post-war period. From the viewpoint of *individual* latecomers, however, the distinction between Abramovitz's realisation and potential factors becomes blurred. Abramovitz's realisation factors, which have mainly to do with international conditions, are in fact the external environment for individual latecomers. They look more like factors governing the potential of a latecomer at a point in time.

On the other hand, it is difficult to regard social capability simply as a potential from the viewpoint of individual countries if changes in social capability are considered. It seems useful to distinguish between *initial* level of social capability at the point in time when catching-up is commenced, and *changes* in it afterwards. From the viewpoint of individual latecomers, initial levels of social capability surely constitute potential for catching-up. But the changes in social capability are hard to treat merely as potential because they are not entirely dependent on initial conditions.

Viewed in this way, we can modify Abramovitz's notions of potential and realisation to tailor them to the analysis of individual latecomers. The potential factors include all the initial conditions of a latecomer, such as technology gap, initial level of social capability, resource endowments, international conditions, and so on. On the other hand, the realisation factors refer to particular ways of organising the catching-up effort, which are related to the change in social capability.

A common limitation of Abramovitz's analysis and the neo-Schumpeterian entry barrier debate is that they attempt to explain catching-up only with the factors affecting the potential. There is no mechanism to connect the potential with the realisation in their analyses. Factors that affect the potential and the realisation in Abramovitz's schema are all those affecting the potential from the viewpoint of individual latecomers. In the neo-Schumpeterian entry barrier debate, only one component of potential, the characteristics of the technology gap, is investigated in isolation. The strength of Gerschenkron's schema lies in the fact that it deals directly with the interaction between the potential and the realisation from the viewpoint of individual latecomers. It has a built-in dynamic. This is the major reason why we take Gerschenkron's schema as the central analytical tool of this work.

# 2

# TECHNOLOGY TRANSFER, TECHNOLOGICAL CAPABILITY AND NATIONAL SYSTEMS OF INNOVATION

In discussing Gerschenkron's view of catching-up in the previous chapter, we have emphasised that differences in the timing of entry into the process of industrialisation are important in understanding the diverse developmental experiences across countries. The focus of this chapter is narrowed to one aspect of being late, that is, the technology gap, with a view to gaining a more solid understanding of the process of late industrialisation through looking closely at this one aspect and its connections with other aspects of being late.

One major reason why the technology gap makes the development of latecomers different from that of forerunners lies in the possibility of technology transfer. Forerunners rely mostly on their own innovating capability in bringing about technological progress. By contrast, technological progress in latecomers begins predominantly with technology transfer from forerunners. If the process of technological progress in forerunners is skewed towards innovation, that of latecomers is skewed towards utilising technology transfer. In a narrow sense, catching-up can be understood as a process of narrowing the technology gap through technology transfer and then shifting the emphasis towards innovation.

Recently, neo-Schumpeterian economists, among others, have made a valuable contribution to deepening our understanding of the process of technology transfer. We shall first summarise their major findings. Then, an attempt will be made to put their findings and interpretations in a broader context. It seems that some interpretative tools employed by the neo-Schumpeterians are not well developed as yet. Particular attention is paid to the concepts of 'technological capability' and 'the national systems of innovation'. The concept of technological capability, we argue, is too broad to guide empirical research or to provide us with a meaningful causal imputation like Abramovitz's concept of social capability. And the concept of the national systems of innovation is, at best, at an exploratory stage. It needs to be complemented by a more in-depth analysis of institutions and institutional change.

## 2.1 TECHNOLOGY TRANSFER AND TECHNOLOGICAL CAPABILITY

The neo-Schumpeterians are a group of economists who premise their economic analyses on technology and technological progress. This is based on their common belief that:

> a number of important economic problems can be powerfully illuminated [by examining technologies] . . . because the specific characteristics of certain technologies have ramifications for economic phenomena that cannot be understood without a close examination of these characteristics.[1]

If the neo-classical economists have treated technology and technological progress largely as exogenous variables, and the Marxist and the Keynesian economists have dealt with them mainly in relation to capital accumulation, the neo-Schumpeterians start their analyses directly from investigating technology and technological progress.[2]

The process of technology transfer has been studied by a number of neo-Schumpeterians in this spirit. One indisputable finding to emerge from their studies is that technology is not transferred from one region to another in its entirety, and that the actual pace of technological progress differs greatly across recipient countries even when the technologies are apparently the same. This finding has brought about a fundamental rethinking of the whole concept of technology transfer. It is far from the simple transfer the term implies. For the technology recipient countries, technologies should be searched, chosen, learned and adapted to local conditions in order to be employed in production. Moreover, they can be improved in the process of learning and assimilation. Thus, Rosenberg (1982: 272) argues:

> the transfer of technology must not be conceived as a once-and for all affair. It is not something that happens at a single point in time. It is, rather, an ongoing activity. Any perspective that ignores this fact is likely to distort the essential issues in technology transfer.

Once technology transfer is understood in this manner, the critical task for latecomers becomes the management of this ongoing process. Technological capability (or capacity) has emerged as an intermediate concept for this management by which the effectiveness of technology transfer is determined. Rosenberg, following the quotation above, argues:

> The successful transplantation of a technology involves the domestic capacity to alter, modify, and adapt in a thousand different ways . . . An economy that lacks the domestic capacity to do these things is most unlikely to make successful use of innovations developed far away and in response to a very different set of circumstances. Conversely,

> an economy that possesses or can acquire this capacity is in a position to draw upon more advanced technologies abroad in ways that can yield spectacular results.

Westphal, Kim and Dahlman (1985) refine the concept through separating it from the elements of technology or technological knowledge. They assert:

> Most important is that a country does not have to possess all the capabilities needed to provide each of the various elements of technology. All the elements of technology can in principle be obtained – individually or in a variety of combinations – from foreign sources.[3]

In this milieu of abundant technological knowledge open to latecomers, technological capability refers to the ability to utilise it. So they define technological capability as 'the ability to make effective use of technological knowledge'.[4] Other studies of technology transfer and technological capability seem to agree with this definition.[5]

Abramovitz's concept of social capability is basically the same as that of technological capability. Both Rosenberg, one of the earliest economists to employ the concept of technological capability, and Abramovitz, who has made the concept of social capability fashionable, have actually borrowed their terms from Rosovsky's account of Japanese economic development.[6] Ohkawa and Rosovsky (1973: 212) define social capability as 'a country's ability to import or engage in technological and organizational progress'. Like technological capability, social capability is regarded as a determinant of the utilisation of imported technologies in latecomers.[7]

It seems important to note that, in the definition above, technological (social) capability is regarded as a *transformer* of the imported technologies into technological progress in latecomers.[8] Here, the imported technologies are inputs and technological progress is an output of the transformer, technological capability. This is very similar to the method employed in the traditional production function analysis, where technology is depicted as a transformer of the inputs such as capital and labour into an output such as physical products, as Figure 2.1 demonstrates. In this respect, technological capability can be understood as a *technological progress function* in the same way as technology is depicted as a production function in the traditional economic analysis.[9]

A vexing problem in analysing technological capability arises from the fact that it is not a fixed transformer. The efficiency of the transformer changes over time. First, technological capability changes as a direct result of technological progress, i.e., owing to the new technological knowledge acquired. Some technological knowledge attained in the process of assimilating specific technologies can be a basis for searching, assimilating

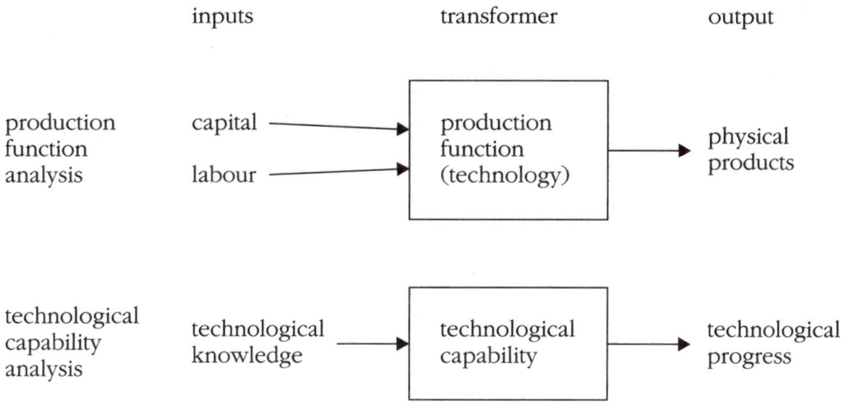


*Figure 2.1* A comparison of production function analysis and technological capability analysis

and improving on new technologies in the same area. For instance, technological knowledge acquired in the operation of an imported textile machine can be used in improving the function of that machine or in selecting another model afterwards. Moreover, some technological knowledge may be more broadly applicable. For instance, the knowledge acquired in the production of TV can be applied to the production of VTR.

Second, what is acquired in the process of technological progress is more than mere technological knowledge. There are components of technological capability which are not exactly technological knowledge itself, although the line separating these non-technological and technological components is often difficult to draw. For instance, in the process of working with the imported technologies, the skills of management needed to organise production or to negotiate with foreign firms can be improved at the level of individual firms. At the level of national economy, the procedure of coordination between public and private sector may be facilitated if government and private firms cooperate in the process of technology transfer. Or the procedure of policy-making and implementation in the government can be enhanced.

There are also both specific and general parts in the non-technological component of technological capability, just as in the technological component of technological capability. Not only can the managerial skills specific to the particular technology assimilated be improved, but also the general managerial skills to organise production may be enhanced. For instance, managerial skills acquired in the process of assimilating TV production technologies can be used in setting up other businesses which are not related to TV technologies. This kind of general managerial capability would be a major source of the diversification process that can be frequently found in late industrialising countries, as will be discussed later (section 3.2.2).

These complex characteristics of technological capability, both specific and general, both technological and non-technological, mainly come from the fact that the essence of technology transfer is learning. Learning is never an automatic process. It is a purposive goal-seeking activity which should be accompanied by some kind of effort. In the process of learning, one acquires not only the specific knowledge originally aimed at, but also knowledge which can be applied to other fields. Moreover, one comes to acquire more general knowledge or competence related to directing the learning effort in future.[10]

Hence, a 'cumulative causation': once technological capability is accumulated over a certain level, it becomes easier to assimilate and improve on imported technologies. Technological capability over a certain level tends to strengthen future capability and vice versa. It is therefore not surprising that most neo-Schumpeterian studies of technological capability or technology transfer stress the strategic importance of the acquisition (creation or accumulation) of technological capability in the latecomers.[11]

However, the interaction mechanism discussed above causes a problem if one wants to analyse the change in technological capability over time. Analytically, this is very similar to that which arises in the neo-classical production function when dynamics are introduced. In the standard neo-classical models, economic growth is conceptually divided into the increase in inputs such as capital and labour, and the increase in the efficiency of the transformer, i.e., technological progress, and economic growth is depicted as the *sum* of these two increases.[12] One of the main criticisms of the neo-classical growth literature has been directed at the difficulty it has in dealing with the actual interaction between the increase in inputs and technological progress.[13] Moreover, the interaction mechanism between technological progress and output growth, which is captured in the so-called Verdoorn's law, is also difficult to analyse using the neo-classical production function.[14]

Likewise, it is often difficult to distinguish the increase of technological knowledge introduced in latecomers from the accumulation of technological capability on the one hand, and, on the other, technological progress from the accumulation of technological capability. In the neo-Schumpeterian literature, the relations between the inputs, the output and the transformer are actually very blurred. Quite often, technological capability is emphasised as a precondition for technological progress in latecomers.[15] In some cases, technological progress in latecomers is identified as the process of the accumulation (acquisition) of technological capability.[16] And it is sometimes argued that technological capability is partly an output of technological progress.[17] If neo-classical economists have attempted to separate those variables, which are not quite separable in reality, in order to attain conceptual simplicity, the neo-Schumpeterians seem to have sacrificed conceptual clarity by not clearly specifying the demarcation lines or the direction of causation among the variables.

We do not deny the importance of what is signified by technological capability in the development of latecomers. Doubtless, the success of technology transfer in a latecomer is critically dependent on its ability to utilise imported technologies. However, the concept of technological capability is still too broad to provide us with a meaningful causal imputation or with a guidance for empirical research. In an extreme case, one could possibly argue that the concept of technological capability subsumes *all* the internal conditions of a latecomer, such as education, firm structure, industrial structure, industrial relations, work ethic, government–business relations, government policies, and so on. The actual meaning of the acquisition (or accumulation) of technological capability is very elusive, because it contains too many components and the interaction mechanism among the components is not clearly established.

In an effort at streamlining the diverse contents of technological capability, it has been pointed out that technological (social) capability can be divided into two components: capability at the level of the individual and at the level of institution. For instance, Westphal, Kim and Dahlman (1985: 171) argue that technological capability 'is the primary attribute of human and institutional capital'. Abramovitz (1986: 222–3) also identifies a country's social capability with technical competence (for which the level of education is a rough proxy) and institutions, as discussed before (section 1.1). Enos (1991: 9–17) divides the contents of technological capability into three: technical knowledge (residing in individuals), institutions (as devices which permit all the requisite technical knowledge, incorporated in human beings, to be assembled and applied) and common purpose (to assure that the outcome of the assembling of skills in an institution should be productive). But, if one accepts the commonly used definition of institution, in which not only actual organisations but also social norms are included, Enos' third component of technological capability can be subsumed into institutions.

But even this categorisation does not seem to mitigate the problems arising from the amorphousness of the concept of technological capability. For instance, as regards the question of acquiring technical competence at the level of the individual, education has been frequently emphasised as a crucial factor by the neo-Schumpeterians and by Abramovitz.[18] If we turn to the originator of the concept of technological capability, however, a different emphasis can be found. Rosovsky points out two major factors behind the expansion of Japanese social capability: the repeated investment spurts and the emergence of new institutions.[19] And he argues that the former brought about the expansion of social capability mainly through 'learning-by-doing'. He points out that 'on-the-job training was also closely related to capital formation' and '[w]orking with new and more advanced machinery and equipment, conceived or manufactured in more advanced countries, raised levels of competence and experience for management and labor, thereby broadening the relevant horizon of technology'.[20]

In effect, Rosovsky emphasises the sequence from importing foreign technologies through investments to enhancing social capability, rather than enhancing social capability *before* importing foreign technologies. This corresponds to Hirschman's (1958) and Rosenberg's (1976) observation of technological progress by 'inducement mechanisms' or 'focusing devices'. The introduction of foreign technologies can create 'scarcities and bottlenecks' which need to be overcome, and can thereby facilitate technological progress. And as we shall emphasise later with the South Korean case (section 8.3), learning-by-doing was an active force in catching-up even in a high-tech industry like the semiconductor industry.[21]

It cannot be denied that both the accumulation of human capital and that of physical capital are all desirable for facilitating technological progress in latecomers. But latecomers are constrained by scarce resources and they urgently need to set priorities for various investments. Only further investigations as to which factor, human capital or learning-by-doing, was a more active force for latecomers or for a specific latecomer, and as to how some countries were successful in developing human capital or sustaining a high rate of physical investment, and so on, would bring about meaningful policy implications. A statement such as 'country A was successful in technological development because it successfully accumulated technological competence at the level of individuals' is 'devoid of content' although it may be 'universally valid', as Weber would argue.[22]

The same can be said about the institutional component of technological capability. Even if one can easily agree on the importance of institutions in acquiring technological capability, questions still remain as to what kind of institutions are needed, how they emerge and so on. The neo-Schumpeterians have attempted to explore the institutional component of technological capability in their studies of the national systems of innovation (henceforth NSI). As shall be discussed in the next section, the studies of the NSI are still at an exploratory stage without developing a commonly accepted framework for undertaking research. This is mainly because institutions function in their context, as we shall elaborate later (Chapter 3). The required patterns or functions of institutions change over time as well. We need to develop a framework to help our enquiry into how to construct institutions compatible with the technological or economic impetus in a particular country.

Viewed as above, the only remaining utility of the concept of technological (social) capability lies in the fact that it is an aggregate concept connecting technical competence at the level of individuals, and institutions, to technological progress or economic growth in latecomers. In general, however, 'no one knows just what it means or how to measure it', as Abramovitz (1986: 222) candidly acknowledges. Its significance as an aggregate concept emerges only in particular historical situations when the levels of technological capability can be separated from those of technology of countries (whether they are measured by productivity or per capita income), as in the case of 'the Golden Age of Capitalism' in Abramovitz's analysis. The notion of technological capability still remains a broad concept, only indicating the importance of internal conditions for latecomers in the process of technology transfer. It is not established as yet how to analyse what is indicated by technological capability.

## 2.2 NATIONAL SYSTEMS OF INNOVATION

The NSI is actually a concept developed by the neo-Schumpeterians for analysing the institutional component of technological capability. Although it originated from studies of both advanced and developing countries, in contrast to the concept of technological capability which originated primarily from the studies of latecomers, it is natural that neo-Schumpeterians' common attention to the actual process of technological progress has led them to recognise the importance of institutions regardless of the country studied. Characteristic features of technological progress detected by neo-Schumpeterians are uncertainties and the importance of learning, both of which are critically determined by institutions.[23] Institutions broadly mark off the boundary of uncertainty perceived by economic agents.[24] The intensity and direction of learning are also affected by the institutional setting where learning takes place.[25]

The NSI can be understood as a concept which has emerged to link institutions at the national level to the technological (or economic) performance of a country, based on the recognition of close relations between technological progress and institutions. Here, however, the consensus among the neo-Schumpeterians ends. They have not reached as yet a clear agreement as to how to analyse the NSI.

Nelson (1988; 1990) and Lundvall (1985; 1988) were initially interested in exploring general features of the system. Nelson focused on explicating general characteristics of technology and on describing the roles played by firms, universities and government in a general way. Lundvall also presented a general account of the NSI by focusing on user–producer interaction. On the other hand, Freeman's (1987; 1988) and Fransman's (1990) analyses were initially directed at explicating the NSI of one country, Japan, and drawing some general implications for other countries. Diverse neo-Schumpeterian studies of the NSI have recently culminated in two major volumes: *National Systems of Innovation: Towards a Theory of Innovation and Interactive Learning*, edited by Lundvall (1992) and

*National Innovation Systems: A Comparative Analysis*, edited by Nelson (1993). The following assessment of the neo-Schumpeterian studies of the NSI is primarily based on these volumes.

What is interesting is that the two volumes show contrasting approaches to the analysis of NSI: if the former is theoretically oriented, the latter is historically oriented. Nelson, though his earlier works on the NSI were directed at exploring general features, now arrives at the conclusion that there are as yet no satisfactory theories or tools upon which comparative studies of the NSI can be based. He emphasises 'the absence of a well-articulated and verified analytic framework linking institutional arrangements to technological and economic performance'.[26] And he expresses a modest aim of his volume, which is a compilation of individual studies of fifteen countries, by saying that 'the orientation of this project has been to carefully describe and compare, and try to understand, rather than to theorize first and then attempt to prove or calibrate the theory'[27] because 'it was apparent that the most interesting features of a country's innovation system varied significantly across countries'.[28]

By contrast, Lundvall criticises the expectation that simple addition to the sample of countries would deepen our understanding: 'The weakness [of Nelson's volume] might be the lack of a common and explicit theoretical basis and the fact that the elements brought into the analysis are idiosyncratic, reflecting the special interests of each author.'[29] And the main emphasis of his volume is placed on an attempt 'to present a theoretical perspective which might be used in case studies and to discuss some of the most important sub-systems in the system of innovation'.[30]

These contrasting approaches of Nelson and Lundvall are reminiscent of the *Methodenstreit*, a methodological debate between the Historical School and the Theoretical School in the late nineteenth century, which will be discussed in detail later (section 3.1.1). Nelson does not deny the need for a theory or a common thread to synthesise case studies. But it is a remote objective at this stage when careful empirical studies have not been sufficiently accumulated. He, like the followers of the Historical School, assumes that the principal task of researchers is to concentrate on descriptions of individual phenomena and compare similarities and differences across countries, before they arrive at a more general understanding.[31] On the other hand, Lundvall, just like the followers of the Theoretical School, emphasises the importance of theory for understanding. He argues that any theory is basically a 'focusing device' which 'brings forward and exposes some aspects of the real world, leaving others in obscurity'.[32] And he proposes 'interactive learning' as his focusing device for understanding the NSI.

It seems that Nelson is open to exactly the same criticisms as the Historical School received, that is, the lack of an explicit common thread to organise case studies in his volume. Although it cannot be denied that an understanding of historical processes can be achieved only by understanding individual cases, neither individual case studies nor reasoned comparisons can proceed without a common thread organising the comparison. And a comparison of individual case studies which have been undertaken without an organising concept is unlikely to produce meaningful results. Let us cite one example. In his conclusion, Nelson asserts, regarding the importance of industrial policy in his sample countries, as follows:

> The picture of government policies supporting industrial innovation that I have been presenting highlights the diversity of such policies and programs, and their *generally* fragmented nature – some supporting research and other activities aimed to help industry in universities or public labs, others connected with defense or space or nuclear power, and still others aimed directly at supporting or protecting certain industries or industry groups. This is what I conclude from the country studies of this project. These studies play down the existence of active coherent industrial policies more broadly. The interpretation they present of the policies of nations widely believed to have them is closer to that of modern day infant industry protection with some R&D subsidy than to a well-structured and thought through general policy.[33]

There are at least two problems in Nelson's interpretation arising from the lack of an organising concept (in other words, arising from not formulating a theoretical question from the start). First, when individual researchers present their case studies, their focus may not be placed on industrial policy. In their broad presentations of the national innovation 'systems', their descriptions of industrial policy tend to be fragmented, giving an impression that industrial policies in the case countries *are* fragmented. Second, as we argued before (section 1.1), the aggregate picture does not necessarily negate the importance of industrial policy in some individual countries. As will be discussed in our case studies of Japan and South Korea (Chapters 6, 7 and 8), industrial policy has been a very active force in these countries. Moreover, the scope and the intensity of industrial policy in these countries have changed over time. Under these circumstances, a mere increase in the number of case studies does not provide a satisfactory answer to the efficacy of industrial policy.

By contrast, Lundvall's approach to the NSI is open to a similar criticism to that the Theoretical School received, that is, the separation of theory from reality, as we shall discuss in the next chapter. We agree with Lundvall's contention that theory is a focusing device and the reliance on a theory is indispensable in undertaking a comparative study. But, in his attempt at developing a theoretical framework, he does not specify the scope for the applicability of his framework.

In fact, Lundvall's original conception of 'interactive learning' or 'learning-by-interaction' was based on his observation of the product innovation process in advanced countries. The significance of the concept decreases in industries or countries where close user–producer relations are not so important. Nelson (1993: 510–11) therefore points out:

> Pharmaceutical companies, strong in Germany and the United States, do not seem generally to have any particular strong supplier connections, international or national. In aircraft production, the producers of components and subcomponents increasingly are located in countries other than that of the system designer and assembler. . . . [I]n small countries or for industries that from their start have been export oriented, the main customer discipline may come from foreign customers.

It is often the case that what are presumed to be common factors across countries lose significance as the number of sample countries increases. And if one tries to avoid this problem by still sticking to common factors across countries, one is obliged either to rely on too broad concepts or to reduce them to certain unambiguous concepts, as can be seen in the case of technological (social) capability (sections 1.1 and 2.1). In comparative institutional analysis in particular, it seems better to make the comparative setting explicit *from the beginning*, because institutions function in their context.[34]

For instance, 'interactive learning' may be a useful organising concept in comparing the NSI of some advanced countries and some industries. In other countries or in other industries, however, domestic user–producer linkage can be substituted for by international user–producer linkage. Other learning mechanisms, say, learning-by-doing, learning from competitors' behaviour, and so on, which can be differentiated from interactive learning, may be more important in other countries or other industries. This would be why Nelson argues that 'it was apparent that the most interesting features of a country's innovation system varied significantly across countries', as quoted above. If a comparison of the NSI is undertaken between a country where the interactive learning is important and another country where, say, learning-by-doing is important, the comparative schema should contain guidance as to why and how learning mechanisms are different across countries. The utility of a comparative schema is critically determined by the comparative setting to which it is applied.

The neo-Schumpeterian studies of the NSI in general do not seem to have succeeded as yet in providing a satisfactory organising concept for directing empirical research. As discussed above, some of them are still engaged in descriptions of similarities and differences of the NSI across countries, whereas some organising concepts are advanced without specifying scope of applicability or limitations. In this respect, it can be said that the analysis of the NSI is still at an exploratory stage. The following chapter is devoted to suggesting an organising concept which seems to partially overcome those limitations.

# 3

# INSTITUTIONS, INSTITUTIONAL CHANGE AND CATCHING-UP

We argued at the end of the previous chapter that the neo-Schumpeterian studies of the NSI are still at an exploratory stage because they have not yet succeeded in presenting a satisfactory analytical framework. The primary aim of this chapter is to investigate some methodological problems involved in comparative institutional analysis and to present an alternative framework that will be used for our comparative study of catching-up. The discussion begins by emphasising the inevitable tension between theory and reality (history). And Weber's methodology will be presented as one promising way to accommodate this tension.

In particular, we shall stress that explanations based on universal propositions have only limited utility because they avoid the tension between theory and reality. A case in point is 'New Institutional Economics', especially its transaction cost economics branch, in its attempt at relating institutions and institutional change to economic performance. We shall argue that the concept of transaction cost is too broad to guide us in explicating causal relations between institutions and economic performance like that of technological (social) capability.

We advance Gerschenkron's 'pattern of industrialisation', briefly discussed in the first chapter, as one framework which might be used to grasp diverse institutional responses in the process of catching-up. It can be understood as an 'ideal type' which is basically a 'tool' to enlarge our historical understanding. Its strengths as well as its weaknesses are discussed. An attempt will be made to extend Gerschenkron's schema beyond its original temporal and spatial limits.

## 3.1 THE TENSION BETWEEN THEORY AND REALITY, AND THE ANALYSIS OF INSTITUTIONS

### 3.1.1 The *Methodenstreit* and ideal type

We have pointed out in the previous chapter that neither simple description nor theoretical construction in itself is enough for understanding diverse social institutions. The diverse reality often evades what a theory tells us, although we cannot reach understanding of the reality without some general concepts. There is a clear tension between theory and reality. How to proceed, then, with social research? This question was a major issue in the *Methodenstreit*, the methodological debate between the German Historical School led by Gustav Schmoller and the Austrian Theoretical School led by Carl Menger in the late nineteenth century.

Neither party denied the need for theory. But they differed in their procedures for constructing a theory and in their perception of its scope. In short, Schmoller's position can be summarised as 'empirical apriorism' towards finding general theories or universal laws:

> Descriptive economics provides the material for general theory. This material is the more complete the more fully the phenomena are described in all their actual characteristics, changes, causes, and consequences . . . [it is necessarily of primary importance] to increase, improve and make more accurate our empirical observation, so that with the aid of the better and more extensive descriptive material of every kind, classifications and definitions can be improved, and typical sequences and their connections, causes, and ramifications may be more clearly understood. If a science from time to time concentrates mainly on description it is by no means in neglect of theory, but to provide a necessary foundation for it.[1]

By contrast, Menger held that there existed a 'dualism' between theory and history:

> The phenomenal world can be regarded from two essentially different points of view: *either* as made up of concrete phenomena in their positions in space and time and in their concrete relations with one another, *or* as forms regularly recurring amid the changes of the concrete phenomena, the understanding of these forms being the object of science. From the former viewpoint study is directed at understanding the concrete, or rather the *individual*, and from the latter the *general* aspect of phenomena. Corresponding to these two main points of view we find two great classes of scientific knowledge, the former of which we may briefly call the individual, the latter the general.[2]

Because of this dualism, Menger perceives the scope of theory more narrowly than Schmoller does. The latter regards a theory as something that includes historical complexities. By contrast, the former regards theory as something separated from historical complexities, dealing only with the *general* aspects of social phenomena.[3]

In the economics profession, the tension between theory and history was not satisfactorily resolved, as the Theoretical School was the clear winning side in the *Methodenstreit*.[4] Schmoller's insistence on accumulating empirical research *before* attaining a theory received heated criticisms from most economists. Schmoller himself, in his later years, also had to rely on theoretical concepts in synthesising his works 'in the same way as do economic theorists'.[5] On the other hand, Menger's dualism has not been contested seriously. Hence the subsequent blossoming of *general economics* which leaves out the historical complexities. Mainstream economics has developed largely isolated from institutional development since then.[6]

Despite the fact that the *Methodenstreit* was initially triggered by a debate among economists,[7] its fruit seems to have been most gainfully reaped by a sociologist, Weber. He clearly recognised the tension between theory and history, and suggested a method to incorporate it in social research.

On the one hand, Weber rejects Menger's dualism between theory and reality. He explicitly states that '[o]ur aim is the understanding of the characteristic uniqueness of the reality in which we move'.[8] For Weber, a theory separated from reality is of no value because it does not have any 'cultural significance'.[9] Theories have their uses only if they help us understand history, which is 'a concatenation of unique events and unrepeatable complexities'.[10]

On the other hand, Weber denies the existence of 'laws' from which historical complexity can be 'deduced'. He points out that what people are actually doing when they are searching for laws is searching for a causal 'imputation'. But universal propositions advanced by those laws have clear limitations in guiding us to reach plausible causal imputations, since they are 'devoid of content' coming from their universal applicability. Weber (1949: 80) argues as follows:

> Laws are important and valuable in the exact natural sciences, in the measure that those sciences are *universally valid*. For the knowledge of historical phenomena in their concreteness, the most general laws, because they are most devoid of content are also the least valuable. The more comprehensive the validity – or scope – of a term, the more it leads us away from the richness of reality since in order to include the common elements of the largest possible number of phenomena, it must necessarily be as abstract as possible and hence *devoid* of content. In the cultural sciences, the knowledge of the universal or general is never valuable in itself.

For Weber, the understanding of cultural significance has to do with understanding 'concrete causal *relationships*', not with finding vague laws.[11]

However, causal relationships in reality are extended to 'infinite multiplicity'. Even in an individual phenomenon, 'an *exhaustive* causal investigation . . . in its full reality is not only practically impossible – it is simply nonsense'.[12] For the human mind cannot contain the infinity. Thus, Weber argues that '[a]ll the analysis of *infinite reality* which *the finite human mind* can conduct rests on the tacit assumption that only a finite portion of this reality constitutes the object of scientific investigation'.[13] We are obliged to 'limit' the extent of the infinite causal chains.[14]

Weber's 'ideal type' is proposed in this context.[15] It is a *preliminary* conceptual construct for research. Weber argues that the construction of ideal types is required *before* research is undertaken because the infinite reality cannot be squared to the human mind as it is. An ideal type is not a final result of research but a 'tool' or a 'heuristic device' to facilitate it. It 'has the significance of a purely ideal *limiting* concept with which the real situation or action is *compared* and surveyed for the explication of certain of its significant components'.[16]

As a preliminary tool, however, an ideal type inevitably contains 'presuppositions'. It is not value-free because 'the very problems which define the objectives of interest are dependent on such presuppositions'.[17] Arbitrariness or one-sidedness is surely involved in this selection process. For instance, 'a phenomenon is "economic" only insofar as and *only* as long as our *interest* is exclusively focused on its constitutive significance in the material struggle for existence'.[18] This arbitrariness or one-sidedness is inevitable in the process of abstraction from the reality. Thus 'this mental construct [ideal type] cannot be found empirically anywhere in reality'.[19] This is why it is 'ideal'.

And Weber argues that, in order to help delineating causal imputations, an ideal type should consist of 'nomological knowledge':

> a *valid* imputation of any individual effect without the application of *'nomological' knowledge* – i.e., the knowledge of recurrent causal sequences – would in general be impossible. Whether a single individual component of a relationship is . . . to be assigned causal responsibility for an effect . . . can in doubtful cases be determined only by estimating the effects which we *generally* expect from it and from the other components of the same complex which are relevant to the explanation.[20]

An ideal type does not by itself reflect the actual causality in the real world. It is only concerned with providing '*adequate* causal relationships expressed in rule and with the application of the category of "objective possibility"'.[21]

Then an ideal type thus constructed is applied to the actual history. The purpose is the 'understanding' of concrete social phenomena by looking at their 'configurations'. According to Giddens (1971: 146), Weber indicates, in describing his objectives in writing *Economy and Society*, that: 'the sociological analysis contained in it performs a task of "very modest preparation" which is necessary to the study of specific historical phenomena' and that '"[i]t is then the concern of history to give a causal explanation of these particular characteristics"'.

This is why Weber's sociology is called 'interpretative sociology'. It is not the 'end', but the 'means' for historical analysis.[22] The understanding of social phenomena is attained and enlarged only through an ongoing process of constructing an ideal type and interpreting social phenomena in line with this.[23] And it is an essential part of understanding to detect the limitations of the ideal type employed. Thus, Weber argues that our knowledge is advanced 'by elucidating the degree of *approximation* to which a particular historical phenomenon can be classified in terms of one or more of these concepts [ideal types]'.[24]

As far as the formal procedure of the ideal type construction, which is based on nomological knowledge and aims at providing a partial picture of the reality, is concerned, Weber's methodology does not appear different from that of theoretical economists. Thus, as Schumpeter (1954: 819) points out, 'as a matter of fact, in the epoch of his ripest thought, M. Weber was not unwilling to declare that . . . he saw no objection of principle to what economic theorists actually did'. But Schumpeter adds, following this quotation, that 'he disagreed with them on what they thought they were doing, that is, on the epistemological interpretation of their procedures'.

This difference is related to the treatment of the tension between theory and history. Among theoretical economists, Menger's dualism between theory and history has been dominant. They argue that the general aspects depicted in theory exist in reality and that they are *the subject* of their research. Thus, they do not feel much uneasiness in remaining in the domain of theoretical reasoning, leaving historical complexities to historians. And general theories abstracted from historical and locational considerations are assumed to be applicable to different societies equally. For instance, when theoretical economists compare the economic performance of different economies, they try to find out whether some general causes for economic growth such as capital accumulation, educational level, the degree of state intervention, the level of R&D investment, and so on, are present sufficiently in one place or another. A crucial assumption in this method is that 'a single cause produces a single effect' across countries.[25]

However, for Weber, theories are essentially 'tools' for delineating historical configurations. General causality implied by an ideal type is purely conceptual. In the actual historical process, a causal relationship established in one place may not be present in another place or another time. In fact, it is quite possible that '[a]n event which is "accidental" from the point of view of a given law can equally well be traced back to its causal antecedents'.[26] And 'it is possible, even probable, that different sets of causes give rise to the same effect'.[27] According to Smelser (1976: 142–5), Weber's main comparative analyses are based on delineating different sets of causes searching for 'distinctive combinations of variables' across societies.[28] Weber's sociology is therefore multifaceted, where several ideal types which have their own tendencies are interconnected.[29] For instance, when he attributes the emergence of capitalism in the West to the Protestant ethic, according to Smelser (1976: 125), he qualifies that 'his interest in the religious origins of capitalism was concerned with only "one side of the causal chain"'. And he argues that 'Western rational bourgeois capitalism arose from *a unique combination of a variety of conditions*, of which the religious ethic was only one.'[30]

Viewed in this way, the differences between theoretical economists and Weber are very similar to those between Abramovitz and Gerschenkron discussed earlier (Chapter 1). Abramovitz attempts to explain the catching-up phenomenon by relying on a universal concept, i.e., social capability. But the term is ambiguous since it is devised to apply to *any* follower country. Although social capability appears a universal cause for catching-up, one is required to lower the level of analysis in order to reach a plausible causal imputation.

By contrast, Gerschenkron's method is directed at explicating different sets of causes across countries which brought about the same effect, i.e., catching-up. He pays attention to the fact that different institutional responses have emerged across different countries. What is important to him in the process of catching-up is the emergence of *functional substitutes* which take different institutional forms. This is because he believes that different institutions can play the same role according to their specific historical configurations. This point will be further elaborated in discussing the relevance of transaction cost economics as a scheme for analysing comparative economic performance across countries in the following section.

### 3.1.2 Transaction cost economics and economic performance

An attempt has been made recently by the so-called 'transaction cost' economists, notably by North (1981; 1990), to relate diverse institutions and institutional change to economic performance across countries. The message of transaction cost economics is that the existence of institutions and their changes can be analysed in terms of transaction cost economising behaviour of individuals.[31] And North maintains that a country's successful economic performance can be attributed to its institutional structure which keeps its transaction costs low.

There has been a long tradition of institutional economists. However, there was no common conceptual framework within which to analyse institutions, although they agreed on the importance of institutions in economic activities and the need for incorporating institutions in economic analysis. What is new about transaction cost economics is its attempt at explaining institutions by means of a universal concept, i.e., transaction cost, which is supposed to be analysed by the tools employed in traditional economics.[32]

In assessing the methodology of transaction cost economics and its application to the analysis of economic performance, it is important to note that the concept of transaction cost basically originated from an attempt to rectify the deficiencies of the concept of production cost in standard neo-classical economic theories. According to Williamson, '[p]retransaction-cost economics [neo-classical economics] takes the organization of economic activity as given and characterizes firms as *production functions* with a motive of profit maximization'[33] whereas 'transaction cost economics maintains . . . that organizational variety arises primarily in the service of *transaction cost economizing*'.[34] North also separates transaction cost from transformation cost, which is almost identical to Williamson's production cost. For both of them, the total cost related to economic activities is the *sum* of production (or transformation) and transaction cost. And transaction cost is defined as costs related with 'frictions',[35] or those involved in 'economic exchange',[36] which do not appear in the standard neo-classical theory.

By definition, the contents of transaction cost are specified as a *residual* of those of production cost. All the costs in the real world except those specified in the neo-classical production function are included in transaction cost.[37] Therefore, Williamson, who is mainly concerned with transaction cost at the level of firms, specifies the contents of transaction cost as the costs related to a contract both *ex ante* and *ex post*.[38] In effect, they are costs which are 'hidden' in the neo-classical production function at the firm level. On the other hand, North, who considers transaction cost mainly at the level of the national economy, specifies the contents of transaction cost as all the costs incurred in 'defining, protecting and enforcing the property rights to goods (the right to use, the right to derive income from the use of, the right to exclude, and the right to exchange)', as compared to the costs incurred in 'transforming the physical attributes of goods (size, weight, color, location, chemical composition, and so forth)'.[39] And he emphasises the importance of transaction cost in the national economy with a calculation that more than 45 per cent of the US national income is currently devoted to transacting activities.[40]

Apart from considering this 'hidden cost', transaction cost economists employ the same rational choice model as the neo-classicals do; what drives their system is cost minimising (or utility maximising) behaviour of individuals.[41] This is the main reason why the tools of neo-classical economics are employed in the analysis of transaction cost economics.

One limitation of employing the concept of transaction cost in comparing economic performance across countries comes from the fact that, even in the rational choice framework, '[t]he objective of the economic agent is not to minimise transaction costs as such, but to minimise the sum of transaction costs and production costs', as Matthews (1986: 906) points out. However, Williamson's theory of industrial organisation is based solely on transaction cost considerations, leaving production cost considerations outside his schema. He seems to think that institutional analysis is concerned only with transaction cost, not with production cost.[42] North also deliberately avoids confronting the problem arising from the sum total characteristics of costs involved in production activities by arguing that changes in transformation cost are basically affected by technological change, whereas changes in transaction cost are influenced by institutional change, although he admits that individuals attempt to minimise the sum of transaction and transformation costs.[43]

Based on the distinction between the determinants of production cost and those of transaction cost as above, Williamson and North advance their analyses by exclusively focusing on the transaction costs of institutions. According to them, transaction cost minimising is important because lower transaction costs result in a better economic performance *if* production (or transformation) costs are kept constant. However, they neglect or play down the neo-Schumpeterian insight that the production function, i.e., technology, is also crucially affected by institutions.[44] Institutions are important because they affect production costs as well as transaction costs. And any economic institution incurs both production and transaction costs. The exclusive focus on transaction costs can be justified only as a preliminary step towards assessing the total costs incurred to economic institutions. In other words, we should ultimately aim at an analysis of the combined effect of production and transaction costs, if we are to reach a meaningful conclusion on the relation between institutions and economic performance.

Even if we focus our attention on transaction cost as a preliminary step towards assessing the total cost of institutions, the concept of transaction cost is still problematic. We emphasised before (sections 1.1 and 2.1) that the concept of technological (social) capability in itself is too broad to help us reach a meaningful causal imputation in the process of technology transfer. Likewise, the determinants of transaction cost are very complex and they are quite varied across countries. The concept of transaction cost in itself is still too vague to be operational in comparing economic performance across countries. The interaction mechanism of the components constituting transaction cost, their connection to economic performance, and so on, should be established.

The method taken by North to make the concept of transaction cost operational is to establish a real-world institution which best *approximates to* the neo-classical world which, by definition, has no friction, i.e., no transaction cost. He then compares other real-world institutions to this *pseudo*-neo-classical institution and explores the reasons why the former differs from the latter.

Following the rational choice tradition, he argues that institutional change takes place as individuals are constantly trying to capture emerging opportunities by a series of voluntary contracts.[45] Hence his description of the process of institutional change:

> The process of institutional change can be described as follows. A change in relative prices leads one or both parties to an exchange, whether it is political or economic, to perceive that either or both could do better with an altered agreement or contract. An attempt will be made to renegotiate the contract. However, because contracts are nested in a hierarchy of rules, the renegotiation may not be possible without restructuring a higher set of rules (or violating some norm of behaviour). In that case, the party that stands to improve his or her bargaining position may very well attempt to devote resources to restructuring the rules at a higher level. In the case of norm of behaviour, a change in relative prices or a change in tastes will lead to its gradual erosion and to its replacement by a different norm. Over time, the rule may be changed or simply be ignored and unenforced. Similarly, a custom or tradition may be gradually eroded and replaced with another.[46]

We do not want to argue with the validity of his contention as such, though we do not agree with it. Our focus is on his methodology. He starts his analysis from an assumption that institutional change, driven by the rational choice of individuals, would lead to an idealised institution *if, and only if*, there is no obstacle. In the real world, however, there are many obstacles that hinder institutional change towards efficiency improvements. North refers to the force behind those obstacles as 'path-dependence'. And he argues that institutional change occurs only *at the margin* due to this path-dependence.[47]

As regards the determinants of path-dependence, North argues that the most important factor is the political 'market' because '[t]he polity specifies and enforces the property rights of the economic marketplace'.[48] And he attempts to delineate the conditions for a political market which best approximates to the neo-classical world. Then, he boldly concludes that '[t]he condition [for political market to approximate to the zero transaction cost model for efficient economic change] is *easily* stated' and that '[t]he institutional structure most favorable to approximating such conditions is a modern democratic society with universal suffrage'.[49] He supports his argument with the comparison between the British–North American path and the Spanish–South American path of economic development.[50] A direct application of this conclusion is:

> When we compare the cost of transacting in a Third World country with that in an advanced industrial economy, the costs per exchange in the former are much greater – sometimes no exchange occurs because costs are so high. The institutional structure in the Third World lacks the formal structure (and enforcement) that underpins efficient markets. However, frequently there will exist in Third World countries informal sectors . . . that attempt to provide a structure for

> exchange. Such structure comes at high costs, however, because the lack of formal property right safeguards restricts activity to personalized exchange systems that can provide self-enforcing types of contracts.[51]

There are a number of theoretical questions unanswered by North in reaching his conclusion. For instance, he does not answer the question as to why an institution which best approximates the neo-classical world is most efficient. According to the 'second-best theorem', a world with one distortion from the neo-classical world is not necessarily more efficient than the one with two or more distortions.[52] North also does not consider the cases where transaction and production costs are interconnected. For instance, the effect of the transacting sector on economic performance cannot be considered without investigating its configuration with other 'productive' sectors. If the relation between the transacting sector and 'productive' sectors is considered, the size of the former itself does not give us any definite efficiency implication. In fact, the relation between the size of service sector, to which the transacting sector is subsumed, and economic performance has not yet been empirically established.[53] Without considering the interaction between transaction and production cost, North (1990: 27) naively argues that, if we '*add*' a theory of production to his theory of transaction cost, 'we can then analyse the role of institutions in the performance of economies'. But the critical issue may be to develop a theory which deals with both production and transaction costs at the same time, rather than to add one theory to the other.

Moreover, for those who do not accept the efficiency of the neo-classical world, North's reasoning does not make sense at all. Here, the size of 'frictions' compared to that in the neo-classical world does not give any efficiency implication. North's argument is based on the assumption that the idealised neo-classical world is the best institution in bringing about long-term *dynamic* efficiency. But the neo-classical world is principally geared at *static* efficiency. So Lucas (1988: 12) points out that, even in the neo-classical framework, there is no guarantee that the pursuit of allocative efficiency will bring about 'growth effects', although it has 'level effects'.[54] North advances his argument without providing plausible criteria for dynamic efficiency.

Considering the many deficiencies in North's argument as above, it is difficult to say that his conclusion on the relation between transaction cost and economic performance is supported by his theoretical reasoning. It seems to us that his conclusion is supported only by his comparison between the British–North American path and the Spanish–South American paths of economic development.

It is not difficult, therefore, to find cases to contradict North's conclusion merely by looking outside his comparative setting. The economic recovery of France after the end of the Second World War was carried out with a sweeping nationalisation, i.e., by limiting individuals' freedom for negotiating contracts.[55] Chang (1994) argues, by employing the very concept of transaction cost, that state intervention in South Korea contributed to reducing rather than increasing transaction costs. Elbaum and Lazonick (1986: 11) attribute one reason for the long-term decline of British industry from the late nineteenth century to the British state's 'reluctance to break from *laissez-faire* traditions'.

The introduction of institutions, or transaction costs, in economic analysis itself does not bring about any definite efficiency implication. Thus, as Matthews (1986: 907) points out, transaction cost economics 'has been invoked in support of both market pessimism and market optimism'. If one wants to know efficiency implications of institutions, the context in which particular institutions under consideration function should be specified. A strong state can either promote or retard economic growth.[56] Vertical and horizontal integration of firms may increase either the synergy effect or rigidities in a national economy.[57] Any institution has its own merits and demerits in itself. And its net efficiency is determined only in a particular context.

How, then, to carry out a comparative institutional analysis of economic performance? What we suggest is to reject the employment of an idealised absolute reference point, such as the neo-classical world or *pseudo*-neo-classical world as can be seen in North's analysis, and to develop a comparative schema by directly abstracting from given comparative settings. This is because institutions function in their context, which is quite diverse across space and time. Only by being explicit about the objectives of the comparison can similar and dissimilar factors between countries be separated. And distinctive institutional factors attributable to performance differences across countries vary with the countries being compared. For instance, in North's comparison of Britain and Spain in the eighteenth century, the reason for British dynamism can be attributed to its freer market economy and more democratic polity. But if the British economy in the late nineteenth century is compared with the more dynamic economies of Germany or the US, British *laissez-faire* may be argued to be a growth-retarding factor.[58] One might also attempt to discover different institutional factors by comparing countries which show similar economic performance as in the case of Gerschenkron's schema.

A comparative schema constructed by abstracting from given comparative settings inevitably has spatial and temporal limits. We have only a *finite* number of countries to be compared within their specific *temporal locations*. This kind of comparative schema certainly has its own limitations. In comparison with the schema based on universal concepts, its scope for applicability is constrained. But it has its own strength in that it accommodates the diversity of institutions to some extent. In the next section, we shall present Gerschenkron's 'pattern of industrialisation' as a schema that can deal with the diversity of institutions since it is abstracted from a limited temporal and spatial 'zone'.

## 3.2 GERSCHENKRON'S PATTERN OF INDUSTRIALISATION: TECHNOLOGY GAP AND INSTITUTIONAL RESPONSES

### 3.2.1 Outline of Gerschenkron's schema

Gerschenkron emphasises that industrialisation processes are quite diverse across countries. If this diversity is hidden behind the concept of social capability or reduced to some general factors like 'educational level' or 'industrial mix', as in the analyses of Abramovitz and others, it is put forward as the main task of investigation in that of Gerschenkron. So he explicitly argues in one of his methodological essays that '[its] purpose rather has been to point out the *great elasticity and variability* in the industrialization processes'.[59]

In analysing these diverse processes, Gerschenkron takes the Weberian position. He acknowledges that any actual occurrence 'can be uniquely related to the coordinates of time and space' as historicists emphasise.[60] But he argues that it is inevitable for a researcher to construct a model or an hypothesis because 'the nature of cognitive processes' is directed at generalisation. So he explicitly refers to Weber's ideal type by arguing that 'what Max Weber described as ideal type was nothing but the method of abstraction and generalization without which scholarly thinking, as any thinking . . . is altogether impossible', and 'whatever the research project, no economic historian will escape the necessity of forming, implicitly or explicitly, consciously or unconsciously, some sorts of "ideal types"'.[61] He also cites from a brain biologist that 'the brain functions by fitting against models',[62] and from Goethe that '[i]t is of highest importance to understand that everything factual is already a theory'.[63]

Due to the inevitability of constructing a model or a hypothesis, Gerschenkron (1968: 42) argues that '[t]he question . . . is not whether in economic history one should or should not abstract, but on which *level* of abstraction one should proceed'. And models and hypotheses have their natural limitations: they are inherently *partial* if juxtaposed against actual occurrences. They are at best explanations or expectations of 'connections in individual cases' but 'no one can prove that they *must* exist'.[64] Thus, 'every serious scholar in testing his model . . . must try to discover the limits of its applicability and will be ready to discard it when those limits have been ascertained'.[65]

As far as the formal procedure of constructing a model or a hypothesis is concerned, Gerschenkron's approach is not so different from that of ordinary theoretical economists, just as Weber's procedure of constructing an ideal type is not much different from that of theoretical economists (section 3.1.1). However, he differs from them in attempting to accommodate the diversity across countries in his analytical framework. This is mainly because he regards the differences across countries as more important than common factors in explaining the process of catching-up. And Gerschenkron opts for an intermediate level of abstraction between universal propositions and descriptive history, and constructs a differentiated system, as discussed before (section 1.3). His schema is a 'paradigm of three countries' derived from nineteenth-century European industrialisation, especially from the experiences of Britain, Germany and Russia. Here, the first industrialisation of Britain provides a reference point, i.e., data for the schema, and it is compared with those of Germany and Russia.

The central concept in his differentiated system is 'the degree of backwardness' on the eve of industrialisation. Organising his observation of European industrialisation in the nineteenth century around this concept, Gerschenkron puts forward the following propositions:

1 The more backward a country's economy, the more likely was its industrialization to start discontinuously as a sudden great spurt proceeding at a relatively high rate of growth of manufacturing output.
2 The more backward a country's economy, the more pronounced was the stress in its industrialization on bigness of both plant and enterprise.
3 The more backward a country's economy, the greater was the stress upon producers' goods as against consumers' goods.
4 The more backward a country's economy, the heavier was the pressure upon the levels of consumption of the population.
5 The more backward a country's economy, the greater was the part played by special institutional factors designed to increase supply of capital to the nascent industries and, in addition, to provide them with less decentralized and better-informed entrepreneurial guidance; the more backward the country, the more pronounced was the coerciveness and comprehensiveness of those factors.
6 The more backward a country, the less likely was its agriculture to play any active role by offering to the growing industries the advantages of an expanding industrial market based in turn on the rising productivity of agricultural labor.[66]

These propositions beg a more 'systematic' presentation because Gerschenkron does not fully elaborate on the relations among them. Let us reformulate them around the fifth one, i.e., the institutional pattern of late industrialisation, because this is most frequently cited. Specifically, Gerschenkron argues that investment banks played a leading role in countries of 'moderate backwardness' like Germany and the state provided the main thrust for late industrialisation in countries of 'extreme backwardness' like Russia, whereas unorganised and autonomous market forces played a major role in British industrialisation.[67] Our question, then, is this: what are the factors behind the correlation between the degree of backwardness and the part played by those 'institutional factors'?[68]

The answer can be explored by starting from Gerschenkron's observation of technology and technology transfer. During the latter half of the nineteenth century, technological progress was most rapid in heavy industry and 'evolution of technology and changing composition of industrial output induced growing capital–output ratios and made for increases in the optimal size of plant'.[69] Here, technology transfer is primarily a matter of capital investment, since technological progress is largely embodied in physical capital. And the resulting catching-up strategy is as follows:

> it was largely by application of the most modern and efficient techniques that backward countries could hope to achieve success, particularly if their industrialization proceeded in the face of competition from the advanced country. The advantages inherent in the use of technologically superior equipment were not counteracted but reinforced by its labor-saving effect.[70]

The second and third propositions are directly derived from this. The stress on producers' goods is a way to exploit the advantages given by technological imperatives of the day. In effect, Gerschenkron detects the process of *unbalanced growth* skewed towards producers' goods sector in latecomers. It is better for latecomers to focus their effort on those sectors which are most promising for catching-up. And, given the technological trend of the day, latecomers could gain advantages in producers' goods sector by establishing bigger 'individual' plants, thereby exploiting increasing economies of scale, although the overall capital investment may be lower than that of the forerunner.[71]

The first and last propositions are also the results of the producers' goods orientation in the latecomers. As Gerschenkron notes in relation to Hirschman's (1958) concepts of 'backward linkage' and 'forward linkage', the industrialisation process of European latecomers was more of a development process based on forward linkage pushed by the producers' goods sector.[72] It was less induced by demand for consumer goods from the agricultural sector, as in the case of the British industrialisation. And less reliance on a gradual increase in market demand and more reliance on 'investment for investment's sake' made the industrialisation process of the latecomers more discontinuous than that of the forerunner.

The fourth and fifth propositions are primarily derived from the need for financing this producers'-goods-oriented industrialisation. Consumption should be sacrificed in order to mobilise scarce and dispersed financial resources in latecomers and also because the investment in producers' goods does not have an immediate consumption-enhancing effect. The major functions of 'institutional factors' were to 'increase supply of capital' and concentrate it on large-scale investments in the producers' goods sector. And the need for applying institutional factors was stronger, the more difficult the capital mobilisation.

In clarifying Gerschenkron's differentiated system, it seems important to divide the contents of 'the degree of backwardness' into two components: the one arising from a different 'timing of entry', and the other from a different degree of backwardness itself. As regards the former, it should be noted that British industrialisation faced a somewhat different technological trend from that of Germany and Russia. During the latter half of the nineteenth century, technological progress was 'more vehement' in the production of capital goods and the direction of technological progress was towards increasing economies of scale, as Gerschenkron (1962: 360) points out. But, during the period of British industrialisation in the eighteenth century and the early nineteenth century, the leading sectors were cotton textiles and machinery industries, where economies of scale were not so important. And as shall be discussed later (section 5.2), the imperative of scale economy was not so strong in the first half of the nineteenth century even in the iron and steel industry. If Germany and Russia had faced the same technological trend in the latter half of the nineteenth century as that which Britain faced in its industrial revolution, the catching-up pattern would have been quite different from Gerschenkron's observation. In fact, as Gerschenkron (1962: 10) notes, all through the nineteenth century, 'England's superiority in cotton textile output was challenged neither by Germany nor by any other country' and '[b]ackward countries (although not the United States) were slow to assimilate production of modern machine tools' where 'very special technological skills are required'. Gerschenkron's pattern emerges only with the 'accidental' technological trend during the latter half of the nineteenth century, as he admits.[73]

The backwardness itself is important in Gerschenkron's schema, because it drives his system. If a country remains backward, the degree of backwardness has nothing to do with the pattern of industrialisation. But the former becomes correlated with the latter because the backward countries attempt to overcome the state of economic backwardness. In Gerschenkron's schema, the 'drive' is catching-up effort on the part of latecomers brought about by international competition among countries.

What should be noted is that, in his schema, this drive takes different forms according to the 'degree' of backwardness. In the case of the British industrialisation, the drive can be said to be a natural rationalisation process based on individual entrepreneurs' self-interest because of its 'zero' degree of backwardness.[74] However, in an extremely backward country like Russia, the need for industrialisation was felt most by the state. It was principally 'the military interest of the state' which forced Russia towards rapid industrialisation. The Russian state had to turn itself into an industrial power in order to assert itself as a military power.[75] In the case of Germany, the situation was something between that of Britain and Russia. It required a special drive because of its relatively backward status compared to that of Britain. But its industrialisation was not solely dictated by the state or by its military motivation because it was only 'moderately' backward.

In Gerschenkron's schema, the emergence of specific institutions and their functions is explained by a combination of the degree of backwardness itself and the technological trend of the day. A major function of institutions was, as we pointed out above, to mobilise resources and to concentrate them on the producers' goods sector on a large scale, because, given the technological trend of the day, this was a promising strategy for catching-up. In an extremely backward country like Russia, that function was carried out by the state relying on its coercive power and budget operations. But, in a moderately backward country like Germany, the role of resource mobilisation was mainly undertaken by investment banks, although the state supported the process. Further elaborations on these different institutional responses are discussed in sections 3.2.2 and 3.2.3.

### 3.2.2 The role of the private sector organisation: investment banks, cartels, *zaibatsu (keiretsu)*, and *chaebol*

Gerschenkron's emphasis on the role of banks in a country of moderate backwardness is mainly drawn from the experience of Germany. He attaches great importance to the role played by the investment bank, arguing that it was 'perhaps the greatest organizational innovation in the economic history of the century'.[76] He tested his hypothesis against the case of Italy and found that it worked to his satisfaction.[77] He also argued that similar roles were played by the banks in other moderately backward European countries like Belgium, France, Austria and Switzerland, even though there were differences between individual countries.

Many historians agree with Gerschenkron in his observation that the German investment banks played a central role in the German take-off of the nineteenth century.[78] But there are still strong reservations in applying Gerschenkron's pattern to other European countries except Belgium.[79] If we consider non-European cases, his pattern becomes more problematic. One test case would be the USA, since it was the only non-European country of moderate backwardness in the nineteenth century, except perhaps Australia. According to Sylla (1991: 58), the role of the banks in the USA was never so prominent as in Germany though it was important in some stages of growth and in some parts of the American economy. In particular, the expansion of the US iron and steel industry in the late nineteenth century was based on internal financing, rather than on bank loans.[80] Moreover, if one tries to extend Gerschenkron's pattern of moderate backwardness to twentieth-century latecomers, it is virtually impossible to find countries fitting the pattern. In a sense, the role of the banks in Gerschenkron's schema can be said to have been historically *unique* to just a few European countries in the nineteenth century like Germany or Belgium.

Does this mean, then, that Gerschenkron's view of late industrialisation under the condition of 'moderate backwardness' loses its significance altogether outside his original comparison of Britain, Germany and Russia? Not entirely so. It seems to us, on the contrary, that a slight modification of his schema would save a large part of its significance and this would be consistent with Gerschenkron's general understanding of historical process.

The first thing that should be noted is that the investment banks were a product of historical evolution and therefore their roles were never perfectly designed or fixed from their inception. Initially, they emerged mainly for the purpose of financing railroad constructions in the 1840s and 1850s in France and Belgium.[81] Then they rapidly expanded their activities to other manufacturing sectors. Crédit Mobilier, the earlier version of investment banks in France was not stable and eventually went bankrupt because of 'its enormously swollen industrial portfolio, which greatly exceeded its capital, and its dependence on favorable developments on the stock exchange for continuation of its activities'.[82] The German investment banks improved on the French version by combining investment banking 'with the short-term activities of commercial banks' and by establishing 'the closest possible relations with industrial enterprises'.[83] Once depression hit the German industries in the 1870s, the German investment banks extended their activities into reorganisation of the industries by forcing cartelisations or mergers. The banks even obtained government approval and support for the industrial projects on behalf of the concerned companies.[84] Moreover, once German industrial firms had grown large enough to be equals of the banks at the turn of the century, their *Interessengemeinschaften* (communities of interest, henceforth IGs) or *Konzernes* began to take over the functions of the investment banks by acting 'as a "credit bank" for members of an allied group' as Chandler (1990: 513) observes.

For countries that industrialised later than Germany, it was not necessary either to repeat this gradual process of evolution or to assign the roles played by the German banks only to their own banks. Institutional learning across countries may shorten the period required in the evolution of institutions. And this learning is never a process of simple transplantation of institutions, rather, it is a process of devising *functional substitutes* for the institutions in other countries. A major source of supposed malfunctioning of Gerschenkron's schema seems to arise when one directly relates this specific institutional form, i.e., the banks, to institutional responses in other countries of moderate backwardness in general. If one begins to think in terms of functional substitutes of the German banks in countries of moderate backwardness, the scope for extending Gerschenkron's schema can be broadened as we shall discuss below.

The second important point is that Gerschenkron's pattern is formulated on a *relative scale of three countries*. 'Moderate' or 'extreme' backwardness are not defined as absolute measures. A country of moderate backwardness in his schema is only relatively backward compared to the forerunner country like Britain, but relatively advanced compared to other later industrialising countries like Russia. It takes a middle position in Gerschenkron's 'three country paradigm'. Therefore, it is essential to make the comparative setting of three countries explicit when one applies Gerschenkron's pattern to other countries or to other periods. Institutional responses in a moderately backward country can be properly understood only in connection with institutions in other more advanced or backward countries.

Then, why did investment banks initiate industrial take-off in Germany? In comparison with Russia, the German credit system was *relatively* well-developed to draw financial resources by itself. Germany was not so backward as to require the state's coercive power to mobilise resources. In comparison with England, however, the German industrial sector was relatively underdeveloped and could not initiate industrial take-off by itself. The German investment banks were an alternative institution which assumed the task for which the relatively underdeveloped German industrial sector was unprepared.

Once Gerschenkron's pattern is viewed in this manner, it is not difficult to find functional substitutes for the German banks in other countries. For instance, *zaibatsu* or *keiretsu* in Japan, or *chaebol* in South Korea performed a similar role to the German banks in the nineteenth century in their industrialisation processes. A general term for these institutions would be 'business group', which can be defined, slightly modifying Leff's (1978) definition, as 'a set of firms which act in different product markets under common entrepreneurial and financial control'. In comparison with the British system based on market transactions of independent producers and financial institutions, a common feature of business groups and the (German) investment banks is the increased coordination through non-market measures within private sector. Business groups coordinate through centralised ownership (*zaibatsu* and *chaebol*) or mutual shareholding (*keiretsu*).[85] The German investment banks coordinated the activities of their customer firms through holding shares and sending their directors to the boards of concerned firms in return for providing loans.

In comparison with market coordination, the non-market coordinating mechanisms of these institutions potentially have some advantages in mobilising and utilising scarce resources in latecomers like capital, entrepreneurial capacity, skilled labour, and so on. A main role of the German investment banks was to mobilise financial resources for long-term investment. And in the general situation of capital market imperfection in latecomers, transfers of capital within business groups may enhance efficiency. As Leff (1978: 672) argues, 'to some extent the groups approximate the functioning of a capital market in the less developed countries'. The investment banks, with their diverse industrial interests, may be in a better position than individual firms to make entrepreneurial decisions. Centralised decision-making at the group level can also save entrepreneurial resources. This is the so-called 'central office' effect of groups because 'the group *structure* itself reduces the amount of entrepreneurial capacity which is required per unit of innovative decision making'.[86]

These advantages are most evident in entering new industries. In the case of Germany, the investment banks perceived the trend of rapid technological progress in producers' goods and concentrated their loans on this sector. Thus German catching-up was most pronounced in new capital-intensive industries like steel, chemicals and electrical engineering. Business groups are also better placed to enter new business than individual firms operating only through market transaction because they are often better than the market at mobilising diverse resources. Here is a cumulative causation, as the diversification of business groups itself becomes a major asset for further diversification. Leff (1978: 672) notes that this advantage 'mitigates downward pressures on rates of return to capital which would otherwise occur if firms were restricted to their existing markets'. And their structure itself is more suitable to sustaining new businesses, especially when they are long-term projects requiring a long time for learning or creating new technologies. Financial resources for sustaining projects may be directly mobilised through the investment banks or within the business groups. Various kinds of indirect financial support from intragroup companies may be provided through purchasing products at a higher-than-market price or supplying inputs at a lower-than-market price. Intragroup transfers of managers and skilled workers often make it easier to solve the problems that emerge in carrying out new projects.[87]

Moreover, the broad diversification provides the investment banks or business groups with a broad information network. This makes it easier for them to identify and locate foreign technologies and new market demand. Especially in the cases of Japan and South Korea, general trading firms, which mainly handle international sales and purchase of raw materials and intermediate goods, have played an important role in gathering market and technological information. This information is easily transmitted within a group. Business groups also have stronger negotiating power when they introduce foreign technologies, which may reduce the transfer prices.[88]

However, these structural advantages exist only *potentially*. There always exists a possibility that the non-market coordination mechanism can turn into a rigidity in the economy. For instance, the German investment banks' interventions in individual firms were often resented by businessmen who wanted to expand their businesses more vigorously.[89] And business groups may obstruct new promising entrants merely by abusing their financial strength. Thus it is necessary to analyse how these potential advantages are realised and how the concomitant disadvantages are repressed in particular countries. Here we need a contextual analysis again to assess the net economic effect of particular institutions.

### 3.2.3 The role of the state

In understanding the role of the state in Gerschenkron's schema, it seems important to note that he distinguishes what he called a 'negative' role of the state which is 'in the nature of creating a suitable framework for industrial development', from 'promoting it directly' which might be called a 'positive' role of the state.[90] And he incorporates only the latter in his schema. By doing so, different institutions, i.e., the British (unorganised) market, the German investment banks and the positive role of the Russian state, are compared as *functional substitutes*. Their common function is to 'increase supply of capital' and to concentrate it on growth sectors, as pointed out earlier (section 3.2.1).

This dichotomy between the negative and positive roles of the state is mainly for the purpose of comparison across countries. As an astute historian, Gerschenkron never belittles the importance of the negative role of the state. He notes the importance of such institutional preconditions in the Russian take-off as the emancipation of the peasants and 'the great judicial and administrative reforms of the sixties [1860s]'.[91] Even in moderately backward countries like France, he argues that:

> [t]o a not inconsiderable extent, the industrial development of France under Napoleon III must be attributed to that determined effort to untie the strait jacket in which weak government and strong vested interests had inclosed the French economy.[92]

He also points out that a similar role was played by the agrarian reforms in Germany.[93]

However, these roles can be carried out only by the state because they are related to the measures 'to remove obstacles that had been earlier created by the state itself'.[94] In other words, there is *no functional substitute* for the negative role of the state. Neither the banks nor the market can be its alternative. In a sense, the negative role of the state is important in any country regardless of the degree of backwardness. This comparability problem would be a main reason why it is not incorporated in Gerschenkron's schema.

Only the positive role of the state can be compared with other substituting institutions, like the market or the investment banks, using the degree of backwardness as an organising concept. With the separation of the positive role of the state from the negative one, Gerschenkron's schema becomes an *economic* historical hypothesis. It treats the negative role of the state, which is more related with political process, as data, leaving it outside the schema.[95] As far as the role of the state is concerned, therefore, Gerschenkron's schema would be read as follows: once the negative role of the state is fulfilled, the need for the positive role is determined by the degree of backwardness.

Then, why is the positive role of the state prominent in extremely backward countries? According to Gerschenkron's observation, the negative role of the state was enough for industrial take-off in moderately backward countries because the investment banks emerged and assumed the role of direct promotion of industries. In Russia, however,

> unlike the case of Western Europe, actions of this [negative] sort did not per se lead to an upsurge of individual activities . . . and for almost a quarter of a century after the emancipation [of peasants] the rate of industrial growth remained relatively low.[96]

This was mainly due to its extreme backwardness:

> The scarcity of capital in Russia was such that no banking system could conceivably succeed in attracting sufficient funds to finance a large-scale industrialization; the standards of honesty in business were so disastrously low, the general distrust of the public so great, that no bank could have hoped to attract even such small capital funds as were available, and no bank could have successfully engaged in long-term credit policies in an economy where fraudulent bankruptcy had been almost elevated to the rank of a general business practice. Supply of capital for the needs of industrialization required the compulsory machinery of the government, which, through its taxation policies, succeeded in directing incomes from consumption to investment.[97]

One might understand this method of coercive resource mobilisation as a manifestation of the traditional mercantilistic intervention by the Russian state which had been evident from the reign of Peter the Great. But Gerschenkron (1962: 132–3; 1970: 115) points out that the role of the Russian state during its take-off period in the 1890s was 'unprecedented' in several respects. On the one hand, the state power was applied in an unprecedentedly large scale in utilising the advantages of the latecomer given by technological progress of the day through unprecedented stress on heavy industry, unprecedented extent of technological borrowing from abroad, unprecedented size of plant and enterprise, and so on. On the other hand, the state itself was unprecedentedly organised for the purpose of industrial push by becoming 'the *sole* organizer and carrier of industrial development'. The major innovation here was the comprehensive use of the state budget for industrialisation which was, in effect, a functional substitute for the German banks. So Gerschenkron (1962: 20) emphasises that '[n]ot only in their origins but also in their effects, the policies pursued by the Russian government in the nineties [1890s] resembled closely those of the banks in Central Europe'.

Owing to the nature and degree of state intervention, an unprecedented result was brought about. Previous industrial pushes in Russia had been followed by economic stagnation once the state withdrew its industrial promotion. But, following the unprecedented intervention of the Russian state during the 1890s, the momentum of growth was maintained. On the industrial basis laid by the 'Witte system' in the 1890s, the subsequent industrial spurt during the period, 1906–14, proceeded even with 'the great withdrawal of the State from promotion of industrialization'. Moreover, the economic scenery of Russia now resembled that of Germany during the 1850s with 'the emergence of the banks as a source of fixed capital'.[98]

Evidently, one limitation of Gerschenkron's schema is that it does not deal with the the 'negative' role of the state in the process of catching-up. Although Gerschenkron (1962: 362) acknowledges this as 'the most important aspect of the problem of limitations', he treats the failure cases related to the negative role of the state largely as exceptions, confining himself to Bulgaria, and possibly to Rumania, in his actual analysis of European industrialisation. However, even in Italy and Austria, which Gerschenkron regards as cases of bank-led catching-up in a moderately backward economy, it has been often pointed out that their states acted as 'obstacles' to industrial growth.[99] In the twentieth century, a larger number of failure cases in catching-up can be attributed to the negative role of the state.[100]

Another limitation of Gerschenkron's schema is that it does not deal with the cases of gradual catching-up. In his schema, the urgency for direct competition with forerunners necessitates that latecomers mobilise their resources and concentrate them on high-growth sectors, and this makes their industrialisation a discontinuous process. But there is a possibility that some countries can proceed with their industrialisation with little urgent need for direct competition with forerunners. A case in point referred to by Gerschenkron is Denmark. It did not undergo a great spurt of industrialisation during the nineteenth century, despite its backwardness. One reason would be its lack of natural resources. But the major reason is that it had 'great opportunities for agricultural improvement that were inherent in the proximity of the English market'.[101] In the latter half of the nineteenth century in Europe, this kind of international specialisation might have been evident only in agriculture. But a pronounced trend all through the twentieth century, especially after the end of the Second World War, was the ever-increasing process of globalisation, which has enlarged the room for international specialisation in industrial manufacturing. This possibility of international specialisation seems to have contributed to making the pattern of industrialisation in the twentieth century rather different from

Gerschenkron's, as shall be discussed later (Chapter 8 and Conclusion).

Despite the limitations discussed above, however, Gerschenkron's schema should not be regarded as a 'falsifiable hypothesis' which can be refuted with an increasing number of 'deviations'. Gerschenkron (1970: 130) argues:

> in trying to set up interpretative models historians do not deal in universal propositions which can never be verified and can only be refuted. We deal in particular or existential propositions. It is the very nature of an historical hypothesis to constitute a set of expectations which yields enlightenment and increases the stock of our empirical knowledge within a spatially and temporally limited zone. To determine the delimitations of that zone does not mean at all a refutation of a hypothesis, but on the contrary its reinforcement as a tool of historical understanding.

Why does 'to determine the delimitations', paradoxically, become 'reinforcement as a tool of historical understanding'? This is because we are obliged to see not only those aspects that fit the tool but also other aspects that do not fit the tool at the same time. When a general theorist faces this situation, he/she tends to choose between ignoring those deviations and refuting the tool, depending on his/her perception of the magnitude of deviations. For him/her, 'to determine the delimitations' can never become a 'reinforcement' of a tool.

For Gerschenkron, however, those deviations are inevitable ones no matter how important they are, simply because his tool is as inherently *partial* as any other ideal type. And his schema is that of intermediate level of abstraction which contains temporal and spatial boundaries in it. It is also designed principally to view economic aspects of late industrialisation through 'one-sided accentuation'. Most of the cases that are supposed not to fit the schema are based on different temporal and spatial boundaries, or on different research interests. In a strictly logical sense, they cannot claim to refute Gerschenkron's schema.

As emphasised earlier (section 3.1.1), it is an essential part of understanding to detect the limitations of an ideal type employed. Our historical understanding is enlarged 'by elucidating the degree of approximation' of the reality in view of nomological knowledge constituting an ideal type. Without an ideal type, we can see neither regularities nor deviations. We come to elucidate deviations just because we look at the reality with an ideal type. In this sense, 'to determine the delimitations' becomes a 'reinforcement of a tool'.

And there are two options in dealing with the deviations. One is to construct a new ideal type or a combination of ideal types to incorporate those deviations into a new system.[102] The other is to explore the deviations without constructing a new ideal type. This procedure can be taken either as a preliminary step towards constructing a new ideal type or simply as a way of enlarging our understanding. Gerschenkron does not attempt to provide complementary or alternative ideal types when his own does not 'work'. He simply acknowledges the limitations of his schema and points out that they can be used for a further exploration.[103]

In our elaboration and extension of Gerschenkron's schema in the subsequent chapters, we do not attempt to construct a complementary ideal type either. We attempt to investigate the recurrent patterns and variations only by changing the temporal and spatial setting of Gerschenkron's original schema. To be specific, among the two components of the degree of backwardness discussed before (section 3.2.1), we shall change only the former, i.e., the 'timing of entry', leaving the latter, i.e., the relative backwardness itself, untouched. And by applying this schema to different sets of countries, we shall examine emerging patterns of industrialisation and their implications. In this process, we can see both economic and non-economic aspects of catching-up although our tool is principally geared at the former.

# Part II

# ON INTERPRETING HISTORY

# 4

# COMPARATIVE SETTINGS

## 4.1 AN OVERVIEW OF THE SCHEMA FOR INTERPRETATION

The following chapters contain the actual analysis of late industrialisation of the three countries, Germany, Japan and South Korea (henceforth Korea), based on Gerschenkron's schema, which has been elaborated and extended in Chapter 3. In comparing the catching-up process of these countries, we narrow our focus one step further by investigating two industries: the iron and steel industry and the semiconductor industry. This is mainly because, we believe, the interaction between technology and institutions is better understood with lower-level industry studies than with higher level aggregate ones, and country-level implications can be drawn more gainfully by looking at the concrete process of technological progress which is quite variegated across sectors. We shall discuss characteristics of technologies and trends of technological progress in the two industries which affected catching-up patterns of our sample countries separately (in Sections 4.2 and 8.1).

By combining the sample countries (or regions) and industries along different periods, we compare the following four sets of catching-up process:

1 the development of the iron and steel industry in two different regions of Prussia, Rhineland–Westphalia and Upper Silesia, during the first half of the nineteenth century (Chapter 5);
2 the development of the iron and steel industries of Germany and Japan from the latter half of the nineteenth century to the early twentieth century (Chapter 6);
3 the development of the iron and steel industries of Japan and Korea during the period after the Second World War (Chapter 7);
4 the development of the semiconductor industries of Japan and Korea during the post-war period (Chapter 8).

The sample countries and industries were chosen partly for their representative characteristics. The three sample countries have in common their spectacular economic performances during their catching-up periods:

Germany was one of the most successful late industrialisers in the nineteenth century whilst Korea was one in the twentieth century, with Japan in between the two. The iron and steel industry has been closely associated with national economic prowess since the Industrial Revolution.[1] Compared to the iron and steel industry, the semiconductor industry is relatively young; it was born in the middle of the 1950s and its impact on other industrial activities was beginning to be seriously felt only from the late 1970s. But it rapidly took the centre stage of industrialisation process as the 'microelectronics (ME) revolution' signifies.

More importantly, however, the choice of sample is geared towards fitting our research objective and tool. We attempt to broaden our understanding of the catching-up process with the degree of backwardness as an organising concept in conjunction with looking at changing technologies. First, we locate some spatial settings to which Gerschenkron's 'three country paradigm' can be applied. As emphasised before (section 3.2), it is necessary to have three countries (or regions) clearly distinguished by different degree of backwardness if one wants to employ Gerschenkron's schema. We then go on to investigate the implications of both the recurrence and the variations of Gerschenkron's original pattern.

In Chapter 6, we attempt only a slight modification of Gerschenkron's original setting by putting Japan in Russia's place. There is no change in temporal setting. By so doing, we can elaborate on the original pattern and see the variations derived from the differences between Russia and Japan. In Chapter 5, we change both spatial and temporal settings. Britain is still the forerunner country. But, two different regions of Prussia, i.e., Rhineland–Westphalia and Upper Silesia, are compared as moderately backward and extremely backward regions. We can see some recurring patterns from Gerschenkron's original schema. There are also variations from the change in temporal setting.

If we move on to the twentieth century, the position of the USA as incontestable technological leader after the end of the Second World War opens a possibility to apply the catching-up paradigm. The large technology gap created by the USA forging ahead provided other countries with a strong impetus for catching-up during the post-war period.[2] One complicating factor in this period is the fact that the process of industrialisation became more globalised than in the previous century. Japan and the West European countries can be regarded as of 'moderate backwardness', since they were only 'technologically, not socially backward' compared to the USA at the beginning of the period, as Abramovitz points out.[3] But there were a number of new participants in the global industrialisation process, broadly designated as 'developing countries'. Even if they might be categorised as 'extremely backward' countries as compared to West European countries and Japan, they entered the industrialisation process with extremely diverse conditions among themselves.

In applying Gerschenkron's schema in this situation, what seems to us important is the interaction with moderately backward countries. East European countries were blocked from interacting with a moderately backward region, i.e., Western Europe by the international politics of the Cold War. Latin American and South Asian developing countries did not have a middle-level country which could act as a crucial intermediary between the frontier country, the USA and themselves. Only East Asian countries like Korea, Taiwan, Singapore, and so on, had Japan as an intermediary between them and the USA. They not only learned a lot from the Japanese experience, which was a recent model case of catching-up, but also received many technologies and institutions that Japan had assimilated, which were more suitable to their needs than the original American ones. Moreover, they also developed facing strong competitive pressure from Japan. In this respect, one might argue that the existence of Japan as a middle-level country contributed to the dynamic growth of the East Asian region.[4]

Among those East Asian countries, Korea has been the greatest beneficiary of the Japanese catching-up as Russia was of the German catching-up in the previous century. Japan was a major source of capital, technology and intermediate materials for Korean industrialisation. And Korea imitated Japanese catching-up strategy in many respects, as shall be discussed later (sections 7.2 and 8.3). Therefore, it becomes plausible to construct a Gerschenkron's 'paradigm of three countries' with the USA, Japan and Korea. They were clearly distinguished by different levels of development while interacting closely with each other.

In Chapter 7, we shall compare the catching-up processes of Japan and Korea by focusing on the iron and steel industry, a typical heavy industry on which Gerschenkron's schema was based. As will be pointed out, the technological trend in the industry was basically the same as that in the latter half of the nineteenth century. But some specificities in Japan and Korea, in conjunction with the increased globalisation in the world economy, brought about their distinctive catching-up patterns. In Chapter 8, we investigate the same issue, concentrating this time on the semiconductor industry, a typical 'new' industry in the twentieth century. This chapter aims to examine differences in catching-up patterns by applying different technological imperatives to the same set of countries.

## 4.2 TECHNOLOGICAL PROGRESS IN THE IRON AND STEEL INDUSTRY

The iron and steel industry is a typical producers' goods industry on which Gerschenkron's schema is based. What should be noted, however, is that the technological trajectory of this industry, moving towards increasing economies of scale and integration of production process as depicted in his schema, was clearly established only from the middle of the nineteenth century. Before then, technological imperatives in the industry were somewhat different from what is commonly conceived at present.

One distinctive technological feature of the industry before the mid-nineteenth century was the separation of the iron-making from the steel-making process. In the iron-making process, technologies had been evolving, though very slowly, towards increasing economies of scale since the invention of the blast furnace around 1500.[5] However, steel-making remained a highly skill-intensive process until the introduction of the Bessemer process in 1856. Thus, many steel shops could be located far away from blast furnaces.

The British forging ahead in the iron and steel industry from the late eighteenth to the late nineteenth century was based on *mass production of iron* by inventing and adopting new methods, coke-smelting and puddling processes, which increased the economies of scale in iron-making significantly. Charcoal had been a principal fuel in iron- and steel-making before the introduction of coke-smelting and the puddling process.[6] In the continental European countries and the USA, charcoal was a dominant fuel in iron- and steel-making even until the middle of the nineteenth century. But transition from the charcoal-based method of iron-making to coal-based took place much earlier in Britain.[7] This transition was a breakthrough towards mass production of iron.[8] By 1848, Britain was smelting more pig-iron than the rest of the world put together.[9]

However, British leadership was not so prominent in steel-making as in iron-making during the first half of the nineteenth century. This was mainly because steel-making was still a highly skill-intensive craft-based process. There had been no tremendous gap in the skills of steel-makers between Britain and her continental counterparts. And the technological gap in steel-making was never big enough to overcome the price differentials resulting from transportation costs and the wage differentials. As shall be discussed later (Chapter 5), the discrepancy between the technological gap in the iron industry and that in the steel industry significantly affected Germany's catching-up pattern during the period.[10]

Germany's overtaking of Britain took place with the beginning of 'the Age of Steel'. The Bessemer process which was introduced in 1856 revolutionised the whole steel-making process. Until then, steel-making was a costly and time-consuming process. Cast iron had to be converted to wrought iron, and then melted or processed in order to be changed to steel (see Figure 4.1). But with the Bessemer process, it became possible to make steel simply by blowing air across molten pig-iron coming out from the blast furnace. It took only ten or twenty minutes to make Bessemer steel as against twenty-four hours for the equivalent amount of traditional steel.[11] With the adoption of Bessemer-style processes, the cost of steel-making dropped by 80 per cent to 90 per cent between the 1860s and the mid-1890s, leading to *mass production of steel*.[12]

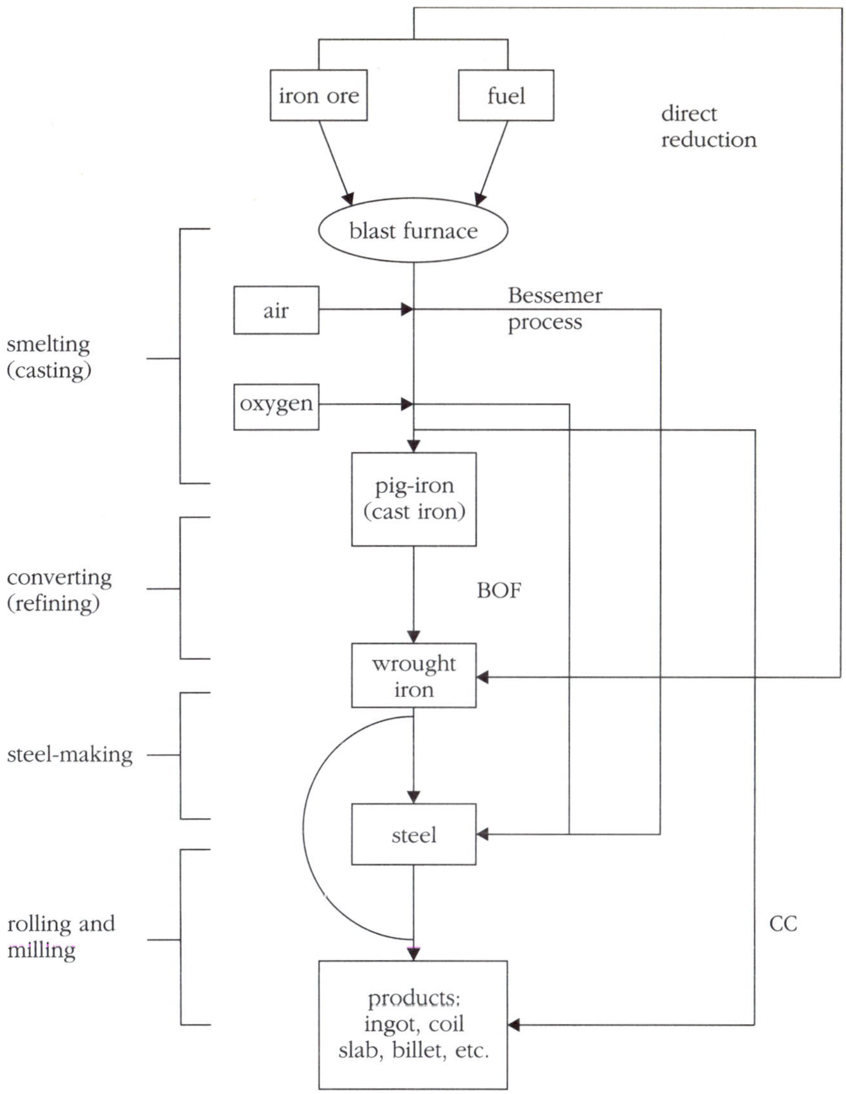


*Figure 4.1* Iron- and steel-making process

Initially, Germany could not adopt the Bessemer process effectively because the original Bessemer process worked only with iron ores which contained a low level of phosphorus whereas most German iron ores contained a high level of phosphorus. This problem was solved only in 1879 by the introduction of the Thomas–Gilchrist basic process developed by two Englishmen. German producers were very quick in adopting the Thomas basic process. During the first half of the 1890s, Germany's production of steel surpassed that of Britain, and by 1903, Germany overtook Britain even in production of pig-iron.[13]

Technological imperatives posed by the Bessemer process were twofold. First, the size of steel mill could be increased dramatically. As mentioned above, steel mills were not integrated with blast furnaces before the introduction of the Bessemer process. But it now became inevitable that iron-making and steel-making processes should be integrated. Second, since the speed of the steel-making process was increased substantially, the problem of process control became acute. Germany exploited and countered these technological imperatives better than Britain did, mainly by establishing 'bigger and bigger' plants and increasing the integration of production (section 6.1).

Along with the German overtaking of Britain, the USA also overtook Britain in steel production around the turn of the nineteenth century. The functional pattern of overtaking was very similar to that of Germany,[14] although the banks did not play a significant role in the USA.[15] The US steel producers dominated world production all through the first half of the twentieth century. Their main competitive edge was largely based on the fast-growing internal market, which was largely insulated from foreign competition. In this protected market they advanced and perfected the Siemens–Martin open-hearth process.[16]

Japan was simply a follower country during the first half of the twentieth century. The initial technology gap between Japan and the West European countries was so large that it took a couple of decades for Japan even to assimilate advanced technologies. But it embarked on Gerschenkronian-type catching-up strategy by introducing state-of-the-art technologies and exploiting increased economies of scale (section 6.2).

After the end of the Second World War, however, Japan soon became the technological leader in the iron and steel industry. It utilised two new technological opportunities. The first was the development of the basic oxygen furnace (BOF), which was invented in Austria in 1952. This is a method of blowing 100 per cent oxygen onto molten pig-iron instead of the ordinary air, which contains only 21 per cent of oxygen. With the help of the BOF, the steel-making time could be reduced to one-tenth of what it had been. Japan took the risk of this unproven technique and succeeded in utilising it.[17]

The second opportunity was the development of continuous casting (CC). This shortened the manufacturing time of slabs and billets by making them directly from the steel-making process rather than by processing ingots. Although CC was invented in America, Japan was quickest in adopting it. Both BOF and CC required an extremely high level of quality control. For this purpose, Japanese steel-makers developed computerised process control and used it far more than any other country.[18]

Japanese steel-makers exploited these technological opportunities by adopting a global production system, that is, by establishing coastal plants, importing raw materials from all over the world and exporting their products to the world market.[19] The size of Japanese newly built mills was increased dramatically, and Japan became the capitalist world's largest steel exporting country in 1973 (section 7.1).

The iron and steel industry of Korea has been dominated by one state enterprise: Pohang Iron and Steel Company (POSCO). From its establishment in 1968, POSCO faithfully followed the Japanese catching-up model without exploiting path-breaking technologies equivalent to the Thomas basic process or the BOF. Gerschenkronian investment strategy is again pronounced in POSCO's rapid catching-up (section 7.2). POSCO was ranked as the second-largest steel company in the capitalist world in 1993 and is often referred to as the most cost competitive steel-maker in the world in low- and medium-quality steel.

# 5

# REGIONAL CONTRAST IN THE DEVELOPMENT OF THE IRON AND STEEL INDUSTRY IN GERMANY, 1800–1850

There is a consensus among historians that Gerschenkron's pattern best fits the German take-off period from the 1840s to the 1870s.[1] Emphasising discontinuities of the industrial take-off, Gerschenkron does not pay enough attention to the gradual and cumulative process of industrialisation prior to the 1840s. However, the impact of the British Industrial Revolution had already begun to be seriously felt by the continental European countries from the late eighteenth century. And, especially with the renewed inflow of British products into the continent after the end of the Napoleonic War in 1815, individual German states and industries, which had been protected by 'the Continental System', had to respond to this challenge. Moreover, although the German states did not reach a political unification until 1871, they could at least begin to form a united front against the competitive pressure from the British products with the movements towards the *Zollverein* (customs union). In this process, 'individual, local examples of rapid industrial growth occurred commonly long before 1840', as Tilly (1991: 178) points out.

The iron and steel industry is a case in point. Until the middle of the nineteenth century, the German iron and steel industry on the whole was predominantly charcoal-based and was the slowest among the West European countries in adopting the coal-based processes developed in Britain.[2] The overall iron production in the German states stagnated compared to their neighbours. For instance, the pig-iron output of Prussia, the strongest economy among the German states, in 1837 was only 96,000 tons while that of France in 1835 was 395,000 tons. Germany's output was also far behind that of Belgium even until the 1850s.[3] But this aggregate picture conceals an important feature of the German iron and steel industry during the period: a sharply contrasting development across regions and by sectors. One region of Prussia, i.e., Upper Silesia was nearly fastest on the continent in installing coke-smelting furnaces successfully. Despite the retardation in pig-iron production, some German iron works in Rhineland–Westphalia demonstrated a high rate of catching-up with their British counterparts in the later stages of iron production like wrought iron- and steel-making.

The investment banks did not emerge until the end of this period. But Gerschenkron's pattern of industrialisation has some explanatory power even in this period. That is, the different degree of backwardness on the eve of the introduction of the modern iron-making process resulted in different institutional responses. In an extremely backward region, i.e., Upper Silesia, the Prussian state was involved in a direct promotion of the industry whereas it played mainly a guiding role in a moderately backward region, i.e., Rhineland–Westphalia.

## 5.1 UPPER SILESIA: A STATE-LED DEVELOPMENT

Upper Silesia was located at the south-eastern edge of Prussia, farthest away from industrial Britain. Traditional textile industries had been prospering in this region; but most of the area was covered with forest and was sparsely populated. The only economic resources favourable to the development of the iron industry were woodlands, deposits of coal, and low-quality ore, although these were not found in as much abundance as in other industrial regions of the continent.[4] As far as the potential of this area was concerned, no one could expect that one of the earliest successful continental emulations of the British Industrial Revolution could take place in this remote region.

However, iron production was beginning to expand from the middle of the eighteenth century. Coke-smelting was introduced in this region as early as in 1791. At first, iron ore was smelted with coke in a small charcoal furnace. Then, blast furnaces specifically to use coke, not in combination with charcoal, were built around the turn of the century. These iron works in Silesia 'were among the first in continental Europe to smelt coke successfully'.[5] In the opening years of the nineteenth century, 'iron technology in Upper Silesia had been amongst the most advanced in continental Europe'.[6] The iron works at Königshütte were referred to as 'one of the best arranged in Germany for smelting iron-ore by means of coke'.[7] Although the iron industry in Upper Silesia stagnated in the 1820s and 1830s, the momentum was regained in the 1840s. The whole industrial complex of Silesia was described in 1842, perhaps with exaggeration, as 'the equal of England and foremost on the continent of Europe'.[8] From 1823 to 1853, Upper Silesia 'produced over 40 per cent of the total Prussian output of pig-iron, and a similar proportion of refined iron'.[9]

The driving force behind the industrial development of Upper Silesia was the political and military ambition of the Prussian state. Upper Silesia was a territory newly acquired by Frederick the Great in 1742 and it was under potential military threat from the Habsburgs, the former possessors of the territory. Moreover, traditional industrial activities based mainly on textile trade with Saxony, Austria and Bohemia were on the brink of disruption by the retaliatory tariffs introduced by the governments of Vienna and Saxony.

For the Prussian state, '[i]t was obviously necessary to integrate the Silesian economy with that of the older Prussian provinces and to encourage the development of new industries in Silesia'.[10]

The Prussian government was particularly interested in exploiting mineral deposits in this area for armaments. A new mining law in 1769 made it clear that mines were under the strict control of the state, especially of the Royal Mining Office.[11] Major obstacles to the industrial development of the region were removed by the state. One difficulty in establishing new industries in this remote and forested area was the shortage of labour. In order to overcome this, the Prussian state, on the one hand, resorted to force by drafting soldiers and extracting forced labour from villages in the vicinity and, on the other hand, provided various incentives to those who migrated into the area.[12]

Moreover, unlike in other areas of Prussia, new iron works in Silesia were established purely as nationalised ones. Although the King preferred to develop the new iron industry by private entrepreneurs or by corporations run jointly by the state and private shareholders, Reden, head of the Royal Mining Office in Silesia, persuaded the King to agree to the establishment of a purely nationalised enterprise. For he 'believed that the State was in a far better position than private persons to finance and administer new industrial undertakings in a relatively remote part of the Kingdom'.[13] The introduction of modern iron-making technologies in Königshütte was led by Reden's state iron works with the help of British ironmasters like William Wilkinson and John Baildon. These state enterprises functioned as 'model factories' for private entrepreneurs. Following the example of the state enterprises, some feudal magnates in Silesia established new iron works, investing their own money.[14]

Despite successful installation and management of state iron works, the development of the modern iron industry in Upper Silesia was an isolated phenomenon in Prussia and did not spread to other regions until the middle of the nineteenth century. There was no coke-smelting blast in the Rhineland until 1840s. Most iron works in Prussia remained small scale and charcoal-based. Moreover, state iron works in Silesia were not cost-effective, though they were most advanced in terms of technologies and organisation.[15] They depended mostly on the state demand until the mid-nineteenth century. And the locational disadvantages of the region were not fully overcome and this became more evident as competing industrial regions to the west began growing from the mid-nineteenth century.[16]

However, the experiment in Upper Silesia is not without significance. As Trebilcock (1981: 28) argues, '[i]t provides perhaps the first European instance of effective substitution for the lack of an extensive commercial class'. The driving force was 'the politico-military ambitions' of the Prussian state, not overall economic development. But the Silesian case at least demonstrates that even a feudal state could initiate and sustain modern industrial development in an extremely backward region by concentrating its ability on a limited region and on limited industries. Gerschenkron's pattern of extreme backwardness is applicable, albeit on a limited scale.

Extensive intervention by the Prussian state has sound explanatory power if the Silesian case is compared with that of Rhineland–Westphalia, which will be discussed later (section 5.2). Unlike in Rhineland–Westphalia, a moderately backward region, the state was the only agent that could establish a new industry in this extremely backward region of Upper Silesia. The same argument was made by Pounds (1959: 196–7) when he compares the Silesian case with similar, but unsuccessful experiments in other German states like Saxony and Thuringia. He attributes their failures to the fact that state intervention was not extensive enough to induce advanced foreign technologies and engineers.[17]

If we compare the Silesian case with other kinds of intervention by the Prussian state, a similar point can be made. The Prussian state was most active among European countries in attempting 'forced industrialisation'. But it is much easier to find unsuccessful cases than successful cases in its mercantilistic attempts. According to Landes (1969: 136), '[m]any of these enterprises failed the moment a change in government personnel or an ideological shift toward *laissez-faire* cut them off from state largesse'. In comparison with those unsuccessful cases, a distinct factor in the development of the iron and steel industry in Upper Silesia was 'the wholehearted support of the state' for a sufficiently long period. The Prussian state's support was sustained from the mid-eighteenth to the mid-nineteenth century. And it was provided in a fuller scale by Reden's pivotal role, as we saw above.

However, the degree of state intervention itself is not sufficient to explain the successful development of the iron and steel industry in Upper Silesia. If we compare the Silesian case with the French one, a different factor emerges. The French government's enthusiasm for a modern iron industry before the Revolution was not in the least weaker than that of its Prussian counterpart. Moreover, the French government took almost the same steps as the Prussian state did when it established modern iron works at Hayanage, Alès, and Le Creusot: sending a government official to Britain to inquire into new technological knowledge, dissemination of it through publications and invitation of a British ironmaster, William Wilkinson, to obtain specifications for installing and managing new furnaces. But the French experiments turned out to be unsuccessful. According to Pounds (1959: 199–200):

> Yet the record of events in France is characterized by chicane and intrigue. Vested interests of individuals stood in the way of the enterprise. In the end, the Le Creusot works were diverted to the production of gewgaws for the royal court. In contrast with the

> industrial development in Prussia, that in eighteenth-century France was grossly mismanaged by a corrupt and inefficient government . . . . The clear-sightedness, ruthlessness, and efficiency of the Prussian government, more than any other factor, explain the early advances in the metal industry in Germany.

It is necessary to enter the area of 'political economy' if we are to probe the question of why the Prussian government was more effective than its French counterpart. But this analysis is outside our schema. We can only point out the fact that extensive state intervention was configured with 'efficiency' of the Prussian government in explaining successful development of the iron industry in Upper Silesia.

## 5.2 RHINELAND–WESTPHALIA: STATE GUIDANCE AND DIFFERENTIAL TARIFF SYSTEM

In contrast to the state-led and isolated development in Upper Silesia, the iron and steel industry in Rhineland–Westphalia was developed predominantly by private initiatives and indigenous skills, and with greater interaction with other regions.[18] The Prussian state did not promote industries of the region by direct measures. One important reason for the contrast in the role of the state can be found in the different degrees of backwardness of the two regions on the eve of industrialisation, as Gerschenkron's schema suggests.

Whilst the iron industry in Silesia began developing only with Frederick the Great's conquest of the region in the mid-eighteenth century, the history of the iron works in Rhineland–Westphalia dated back to the Middle Ages. The iron and steel industry of the region had developed on the basis of traditional regional advantages such as the existence of iron ore deposits, woods and water power. A high degree of skills had been accumulated already through centuries. The division of labour in iron-making within the region was already well advanced.[19] Moreover, industries in Rhineland–Westphalia had been exposed to constant interaction with more advanced neighbouring countries like France and Belgium, as well as with Britain. Under these circumstances, the need for the state's direct promotion of the industry was much less in Rhineland–Westphalia than in Upper Silesia.

During the first half of the nineteenth century in Rhineland–Westphalia, however, there was no functional substitute for the investment banks, i.e., private sector institutions which would have provided 'organisational direction' for catching-up. Nevertheless, a gradual import-substitution process took place in the iron and steel industry. The volume of the iron production in Germany increased continuously and the production gap with its neighbouring forerunners like France and Belgium was narrowed significantly by the end of the period.[20]

In understanding the forces behind this deviation from the Gerschenkronian pattern, it is important to note that technological imperatives in the iron and steel industry during this period were quite different from those in 'the Age of Steel' which began with the introduction of the Bessemer process. In technological catching-up in the Age of the Steel, it was imperative to build integrated steel mills, which were very capital-intensive (section 6.1). Before the advent of the Age of Steel, however, the iron- and steel-making process was neither tightly integrated nor highly capital-intensive. Steel could be produced in a small workshop far away from iron-making furnaces because steel-making was mainly a craft-based process. Moreover, the competitiveness of the old charcoal-based iron-making technologies was maintained until the middle of the nineteenth century. Mainly due to the existence of a still abundant and cheap wood supply on the continent,[21] and partly due to continuous improvement in the old technologies, traditional charcoal-based iron products could compete with the coal-based products for a long time, even after the charcoal-based production ceased to exist in Britain.[22]

Owing to these technological reasons, the need for capital mobilisation did not arise seriously in the private iron works of Rhineland–Westphalia. If one was to initiate catching-up process by building a modern coke-smelting furnace, the problem of financing would have been acute. But there still remained another route for catching-up: a gradual catching-up by relying initially on traditional technologies and skills, and by introducing modern technologies slowly. In fact, the development of the iron industry in Rhineland–Westphalia, as it was in most other continental areas and in the USA, was 'two-pronged': 'On the one side, there was the introduction and considerable diffusion of the new mineral-using techniques; on the other, there was an expansion of old-type plant, improved to some degree, but obsolescent'.[23] In this process of two-pronged development, private iron works in Rhineland–Westphalia could remain invariably small in scale without requiring large-scale investment until the 1840s.[24]

This route for catching-up became possible due to the combination of the previous development of the private sector, i.e., moderate backwardness of the region and the technological imperatives in the iron and steel industry during the period. This option was available neither in Upper Silesia simply because there was no previously developed private sector although the two regions faced the same technological imperatives, nor in the Age of Steel because the technological trajectory in the industry was established towards increasing economies of scale and integration as shall be discussed later (Chapters 6 and 7).

However, the combination of the moderate backwardness and the technological imperatives of the period does not provide a sufficient answer for the competitiveness of the iron and steel industry in Rhineland–Westphalia in comparison with its forerunner, i.e., the British iron producers. For, as Fremdling (1991a: 57, fn. 14) points out, 'Britain undoubtedly was the lowest cost producer during the period in question [1820s–1860s] and hence possessed absolute advantages in producing primary iron products'. There should have been something to compensate the cost differentials between the British and the Rhineland–Westphalian producers in order that the gradual import-substituting process could occur. Here lies the importance of the role of the state in this region. Import duties imposed by the Prussian state were a primary factor which compensated the cost differentials for the German producers, along with transportation costs.

The essence of the German tariff system during the first half of the nineteenth century was differential treatment across commodities within a generally liberal trade regime.[25] In the iron industry, this differential tariff system was applied between the import of pig-iron and that of wrought iron. The Maassen tariff of Prussia in 1818, which provided the basis for the *Zollverein* tariff, treated pig-iron as a raw material and therefore admitted free of duty while around 10 per cent *ad valorem* duty was charged to wrought iron. The *Zollverein* tariff in 1834 increased duty on wrought iron to around 40 per cent, leaving free import of pig-iron untouched.[26]

The effect of this differential tariff system would be obvious. German pig-iron producers were clearly the main losers in this system owing to free import of British and Belgian pig-iron. Traditional charcoal-based pig-iron production almost stayed still during the period while imports were increased substantially. Moreover, the spread of coke-smelting in Germany was retarded by free imports of foreign pig-iron. Thus, Crouzet (1972: 123) points out that 'it was only after the duties on pig-iron and wrought iron had been raised in 1844 that coke-smelting progressed in a big way'.[27]

However, on the whole, the negative effect of the liberal tariff system on the pig-iron producers seems to have been more than compensated by its beneficial effect on the later-stage producers of wrought iron, steel or finished iron products. What should be remembered here is that, in terms of technological level, the German iron and steel industry during the first-half of the nineteenth century was behind its British counterparts only in smelting and refining. As far as steel-making, hardware production and cutlery were concerned, German producers were never far behind the British.

Free import of cheap British pig-iron was in fact a corroborating factor for the competitiveness of the German steel and finished iron producers. Combining existing processing skills with cheap imported pig-irons,

> the most flourishing sector was the secondary metal industries, such as the hardware and cutlery of the Sauerland; and the greatest advance was made in the manufacture of those finished goods – steel and steel products – that demanded special skills and high inputs of labour.[28]

Although the German steel producers were initially retarded in making crucible steel, they began to lead technological progress from the mid-century.[29] And, in the 1840s, '[s]everal of the newer forges [for puddling and finishing] of the Ruhr and Sauerland were among the most modern on the Continent'.[30]

The immediate impact of the differential tariff rates on the refining sector was rather ambiguous. Although free import of pig-iron should have been beneficial to the refining sector, the import of the competitive British wrought iron, on which only 10 per cent of duty was imposed, had a potentially detrimental effect on the German refining sector.[31] The impact of the *Zollverein* tariff in 1834 was more clear. Heavy duties on wrought iron had a clear aim: import-substitution of rails, most of which were imported previously. As Fremdling points out, the differential *Zollverein* tariff system pushed 'the bar iron production of the 1830s forward to a more highly developed technological level than pig-iron production' (1977: 595), and 'Britain increasingly began to lose her *absolute advantage* in the second stage of iron production from the 1840s onwards' (1991a: 49; emphasis added).[32]

The *Zollverein* tariff was revised again in 1844 with the introduction of heavy duties on pig-iron and proportional increase of duties on wrought iron. Duties on pig-iron were 29 per cent *ad valorem* in 1845 and was increased to 50 per cent in 1851. With this revised tariff, a rapid import-substitution took place. In 1843, 90 per cent of rails were imported, in 1853, roughly half of the stock of rails on Prussian railways was of British origin, and in 1863, it was reduced to only 10 per cent. Coke-smelting also prospered under the wall of protection. There was only one coke-smelting work in the Ruhr area before the introduction of the revised *Zollverein* tariff. But by 1850, the number increased to twenty-three. The heavy duties were gradually reduced from 1850 as import-substitution progressed and were finally abolished in 1873.[33]

What is important to note is the *reversed sequence* in the introduction of modern iron-making technologies. In Britain, the puddling process was introduced long after the coke-smelting process.[34] In Rhineland–Westphalia, by contrast, the puddling process was introduced in 1823 but coke-smelting began to be used regularly only from 1849. In Maurice Levy-Leboyer's terms, if the British iron industry developed 'downstream' or 'forwards', the iron industry in Rhineland–Westphalia developed 'upstream' or 'backwards'.[35]

This reversed sequence has some definite advantages over the original sequence in Britain. Technologically, it is 'the normal sequence', as Landes (1969: 175–6) points out. Coke-smelting is a technologically more difficult process than puddling because direct contact between coke and ore in the former complicates the problem of chemical composition of the material employed. Once both coke-smelting and puddling technologies are available, the latter is easier to learn and cheaper to introduce.

Moreover, by reversing the sequence, the costs incurred in the transition from old to new technologies could be reduced by 'technology mix'. At first, traditional charcoal refineries used both domestic charcoal pig-iron and imported coke pig-iron because 'they were effectively interchangeable' except for railway iron.[36] Then they adopted the puddling process gradually. Even new puddling mills were established for technology mix and made a gradual transition. As Fremdling (1991a: 54) observes, 'German puddling and rolling mills first had to learn how to use imported coke pig-iron in combination with, and finally instead of, indigenous charcoal pig-iron'.

It was not an available option for the British wrought iron sector to import cheap foreign pig-iron simply because it was a forerunner. By contrast, this option was widely open to the iron and steel industry in Rhineland–Westphalia because of its 'being late' as well as its geographical proximity to the advanced industrial countries. Moreover, iron works in the region already retained a high level of skills in processing and finishing. By reversing the sequence, a cheaper process of catching-up in the wrought iron sector could take place in Rhineland–Westphalia, firmly based on indigenous skills and with a small amount of capital. The iron and steel producers in Rhineland–Westphalia could pursue their catching-up process mainly by increasing the efficiency of existing capital, rather than by raising new capital.[37] This advantage as a late-comer was not so evident until the demand from the railway sector fuelled the production of domestic wrought iron in the 1840s. But, with already accumulated technological skills and organisations combined with the option of importing cheap pig-iron, the wrought iron sector in Rhineland–Westphalia was 'able to respond quickly to railroad needs'.[38]

Apart from the adjustment of the tariff system, there were several beneficial roles played by the Prussian state in fostering the development of the iron and steel industry of the Rhineland–Westphalian region. The Prussian Reform in the early nineteenth century provided a broad framework for industrial development. In particular, the establishment of educational systems during the Reform period has been cited frequently as one of the most important contribution of the Prussian state to its industrial development.[39]

However, the role of the state in Rhineland–Westphalia was not extended to a direct mobilisation of resources. Major efforts of the state were devoted to fiscal reforms in the region and improvement of infrastructure, especially building new roads and canals.[40] The state did not direct the economy though private firms and mines were officially under the supervision of the state according to Prussia's mercantile tradition. And most economic polices were decided by close consultation with local businessmen.[41] This limited role of the state can be attributed to the moderate backwardness of the region in conjunction with the technological imperatives of the period, as discussed above.

What is interesting is the fact that the role of the Prussian state described above was also interconnected with *political* necessities in the region during the period. That is, the moderate backwardness of Rhineland–Westphalia was closely related to the liberal attitude of the business community which tended to resist the state's direct intervention. This contrasts with the situation in Upper Silesia where its extreme backwardness was related to the absence of a comparable private sector.

The Rhineland–Westphalian region had long suffered from an extreme political division.[42] Prussian territories in the region had been 'acquired piecemeal over a period of two hundred years' whereas 'Silesia was seized in a single campaign and was successfully defended against Austrian attempt to recover the province'.[43] Industries in Rhineland–Westphalia had not developed under the tutelage of a single political entity. Most iron works had developed mainly in peasants' houses under the supervision of their noblemen, and the owners had to manage their iron works in the face of frequently changing and often capricious rulers of the region.[44]

In particular, Rhenish businessmen on the left bank of the Rhine were greatly influenced by liberalism during the period of the French occupation from 1794 to 1815. Previous feudal obligations were abolished. Authority and administrative procedures were rationalised with the introduction of the French legal codes and the French system of local government. Many businessmen participated in local government and could thereby represent their own interests. They remained suspicious about the Prussian state after reintroduction of Prussian rule with the end of the Napoleonic War. The so-called 'Rhenish particularism' was no more than an expression of the attempt at preserving local interests against the still uncertain rulers of the region.

In this situation, it would have been counter-productive if the Prussian state had attempted to intervene in the industry directly. Moreover, the state was bound by its own fiscal difficulty during the period. One of the most urgent problems of the Prussian state was to get out of the debts left over from the Napoleonic War. Fiscal reforms during the period, including the introduction of new tariff systems, were aimed mainly at balancing the budget. The state did not have the necessary funds for investment in industrial projects.[45]

It should be noted that the Prussian state was no less eager to promote industries in Rhineland–Westphalia than in Silesia. In fact, one can easily find the state's eagerness for industrial development of the region in several government documents.[46] But different regional conditions necessitated a different mode of state intervention in the economy. The task of the development of Rhineland–Westphalia was carried out by liberal officials like Stein, Vicke and Sack, whereas that of Upper Silesia was done by a more interventionist official, Reden. The appointment of liberal officials itself contributed to reducing suspicions on the part of the Rhenish businessmen. The Prussian officials also tried to reconcile the differences between Rhenish and Prussian institutions while promising to promote the trade and industry in the region. This reconciliation was also undertaken very gradually.[47] In fact, the Prussian state allowed Rhineland–Westphalia relative autonomy, aiming at a smooth transition of the region into a fully integrated part of Prussia.[48] In this political situation, the adjustment of the tariff system was likely to be more effective than direct state intervention.

## 5.3. CONCLUDING OBSERVATIONS

The regional contrast between Upper Silesia and Rhineland–Westphalia discussed above sheds a mixed light on the applicability of Gerschenkron's schema. On the one hand, it roughly fits Gerschenkron's pattern in the sense that the role of the state was more direct in a more backward region, Upper Silesia, than in a less backward region, Rhineland–Westphalia. And this difference was correlated with the degree of backwardness. On the other hand, however, the role of the private sector in Rhineland–Westphalia does not fit Gerschenkron's pattern of moderate backwardness. There was no conspicuous private sector institution initiating catching-up. A functional substitute for the investment banks in Gerschenkron's schema did not emerge.

This deviation from Gerschenkron's original pattern emerged mainly because the functions of investment banks were not needed due to the different technological imperatives of the day. Catching-up in Rhineland–Westphalia during the first half of the nineteenth century could proceed without a large-scale capital mobilisation. In this situation, adjustments of tariff rates by the state and its guiding role were enough for private producers to undertake their catching-up process. Hence no 'big spurt'. Rather, catching-up proceeded with gradual import-substitution under the guidance of the differential tariff system. A similar process of catching-up can be found in the development of the Korean electronics industry as shall be discussed later (section 8.3). And we shall also discuss later the relation between this gradual process of catching-up and the degree of backwardness in a broader view (section 8.4 and Conclusion).

The role of the Prussian state can be better understood if our explanation above is complemented by 'political economy' considerations. In explaining the role of the state in Rhineland–Westphalia, we pointed out that the moderate backwardness was combined with political necessities in the region. But further research is required to answer why the private sector did not resist the differential tariff system, how the Prussian state could mitigate the possible tensions, and so on. Likewise, our explanation of the role of the state in Upper Silesia needs to be complemented by investigating its political economy and by comparing it with those in other countries or regions.

# 6

# UK–GERMANY–JAPAN COMPARISON IN THE IRON AND STEEL INDUSTRY, 1850–1930

Coal-based iron production became dominant in the 1850s with railway booms in the West European continental countries.[1] Traditional charcoal-based production was almost completely replaced by coal-based production, mainly because railway booms demanded irons in previously unknown quantities, though not in higher quality. Although Britain still remained the largest and the most cost-efficient producer in the international market, the technological gap between Britain and Germany had been reduced substantially. With the enhanced technological capability of the German producers, the Bessemer process 'spread in Germany without considerable time lag from Britain'.[2] And once the phosphorous problem in the Bessemer process was solved by the introduction of the Thomas basic process, Germany overtook Britain rapidly (section 4.2).

The most important institution in this catching-up and overtaking process was the investment banks. Their role in the German industrial take-off from the 1840s has already been emphasised (section 3.2.1). One purpose of this chapter is to elucidate the role of the investment banks by focusing on a specific industry, i.e., the iron and steel industry. But the role of the state cannot be ignored, though it was not extended to a direct mobilisation of resources. The respective roles of the investment banks and the state in the German iron and steel industry will be assessed.

In Gerschenkron's original schema, German catching-up is compared with that of Russia. We replace Russia with Japan in an attempt to extend his schema. The two countries were similar in their degrees of backwardness in the late nineteenth century. Both of them were relatively backward compared to Germany but relatively advanced compared to other countries outside Western Europe. It is estimated that in 1870 their real gross national product per capita was about one-quarter to one-third of that of the UK, with Japan at the lower and Russia at the upper level of this range.[3]

Due to this similarity, the two countries showed a very similar institutional pattern of catching-up: their modern iron and steel industries were developed by direct promotion from the state. However, the Japanese catching-up speed was slower than that of Russia. The Japanese iron and steel industry reached the technological level of advanced countries only in the 1920s, whereas the Russian catching-up was already evident in the early years of the twentieth century.[4] An elaboration of this difference in the effectiveness of state intervention is another task of this chapter.

## 6.1 GERMAN OVERTAKING OF THE UK: THE ROLE OF THE INVESTMENT BANKS AND CARTELS

In section 5.2, we emphasised that the Prussian state's involvement in the iron and steel industry of Rhineland–Westphalia during the first half of the nineteenth century was not extended to a direct promotion because of its fiscal and political constraints, on the one hand, and moderate backwardness of the region, on the other. From the middle of the century, the configuration between the state and its economy began to take a different form. The previous constraints on the Prussian state were weakened substantially. It no longer faced the budget deficit problem. There also emerged a new alliance between industrialists and *Junkers* after the failed revolution in 1848, which should have acted for increasing the state's power in supporting industry.

At the same time, however, the economic backwardness of Germany had been diminished. The growing German private sector increasingly felt many of the existing state's involvements in the economy as an obstacle rather than an inducement for industrial development. Private industrialists began to 'assert their right to manage their business affairs without any guidance from paternalistic governments', which was reflected in the demand for constitutional reform during the revolution period.[5] The overall consequence was the decreasing involvement of the state in the economy.[6] For instance, due to the pressure from industrialists in the Ruhr, the Prussian mining laws underwent a fundamental reform in the 1850s.[7]

The German states did not provide any particular promotional measure for the emergence and expansion of the investment banks. Indeed, the investment banks prospered initially *in spite of* the constraints imposed by the state. The Prussian state viewed them as 'a special object of suspicion' because of their speculative activities as already seen in France.[8] Thus, the first German investment bank, the bank of Darmstädt, could not find its founding place within the Prussian territory. It had to be established in a small German state, Hesse–Darmstädt, adjacent to Prussia.[9] The Prussian state refused to charter banks as joint stock companies until the early 1870s.[10]

However, the Prussian state's initial antagonism towards the investment banks could not restrain their expansion. For it was impossible to block the operation of the investment banks situated outside the Prussian territory from dealing with Prussian industrialists. Moreover, due to the competitive pressure by these investment banks, there emerged *pseudo*-investment banks within the Prussian territory. That is, some Prussian commercial banks, notably *Kommandit* banks, in effect acted as investment banks from the mid-1850s. They were '*Kreditbanken* [investment banks] in *Kommandit* form', as Tilly (1967: 163) points out.

### *The engine of expansion*

The role of the investment banks in the development of the German iron and steel industry can be traced in two ways: direct and indirect. The investment banks indirectly gave a boost to the iron and steel industry by acting as a leading financier for the railway boom. More than three-quarters of the total capital invested in joint stock companies, between 1851 and 1870, went into railway construction.[11] The railway was the leading sector for the German industrial take-off, and the rapid increase of the iron and steel production in the 1850s and 1860s was propelled by the demand from the railway sector.[12]

The investment banks also directly contributed to the growth of the German iron and steel industry as its financier and organiser. Let us first focus on the former role, keeping in mind that it is closely related with the latter. The banks' direct financing became pronounced in the iron and steel industry from the 1850s because of the rapidly increasing capital-intensity with the wide adoption of coal-based production and the arrival of 'the Age of Steel'. Most of the newly established firms in the 1850s and 1860s like Rheinische Stahlwerke, Thyssen & Co., Eisen-und Stahlwerke Hoesche, and so on, were heavily dependent on the banks for their initial capital, though some existing steel-makers like Krupps were initially not entirely dependent on the banks. Cologne's Schaaffhausen'scher Bankverein was, for instance, referred to as 'the financial godfather' of the leading iron and steel firms in the Ruhr.[13]

The Great Depression from 1873 to 1879 further increased the industry's dependence on the banks because survival of many companies was crucially dependent on the banks' 'readiness or lack of readiness to grant credit', only those supported by major banks being able to survive during the period.[14] And the introduction of the Thomas basic process in 1879 provided a further impetus for bank financing in the industry. The price differential between Bessemer and Thomas steel was so great in Germany that even the newly established Bessemer plants had to be converted to Thomas basic plants. But, at the end of the Depression, most producers lacked funds for the new investments required. Moreover, the new opportunities opened by the Thomas basic process attracted several new entrants. The investment banks were almost the sole provider for large-scale financial resources in this situation.[15]

The important point is that this ever-increasing dependence on the bank financing was the *engine of expansion*. Initially, the German producers were not so effective as their British counterparts in exploiting the increased scale economy brought about by the Bessemer process, even with the help of the investment banks. According to Landes (1969: 219), in 1870, '[t]he *most powerful* furnaces in the Ruhr yielded around 250 tons of pig a week' while 'the *average* British unit did almost as much'.[16] But, at this time, the main problem lay in the incompatibility of the original Bessemer process and German iron ore, not in their technological level or institutional structure.

Once this technological problem was solved with the introduction of the Thomas basic process, the advantage of close relation between the banks and the German producers became evident. The German iron and steel industry could build bigger and bigger plants backed by their investment banks. Landes (1969: 263) points out that,

> [a]round the turn of the century, the *biggest* British mills were turning out only as much as the *average* Westphalian works . . . . [and] new British plants in the 1890s were a quarter to a third of the size of their German competitors.

Germany overtook Britain even in the size of the open-hearth plants in which the British producers were specialised.[17] The German producers also countered the process control problem by increasing integration and mechanisation far better than their British counterpart.[18]

The role of the state cannot be neglected in this expansionary process. The German state gave a crucial support to the banks' credit expansion with its 'easy money' policy from the 1870s. As Tilly (1991: 183) points out,

> from 1876 the powerful support universal banks gave German industry and trade rested in part on the payments network and liquidity guarantee provided by the Reichbank. . . . [M]ost important . . . was the virtually unlimited access to the Reichbank's discounting facilities which the credit banks came to enjoy. The German banks could get by with less liquidity and 'lend to the hilt' if demands warranted, because the Reichbank provided extremely liberal rediscounting facilities. In fact, they were so liberal – though not cheap – that bills of exchange held by the banks or their acceptances could be seen as substitutes for central bank notes. . . . Its [Reichbank's] contribution to the 'German system' lay in the flexibility/elasticity of its rediscount policy, especially in times of 'liquidity pressure'. The universal banks, in consequence, could hold more risky portfolios than would have otherwise been the case.

It should be noted, however, that the German state did not initiate this credit expansion. The Prussian state had in fact remained restrictive in expanding its circulation of banknotes before 1870, at least officially. Thus, it is estimated that, in the 1850s, the circulation of foreign banknotes in Prussia exceeded that of the Prussian Bank.[19] But there emerged several 'money substitutes' created by private banks, which eased monetary constraints in the industry.[20] Also, the Prussian state had to expand the circulation of its banknotes 'as a direct response to the competitive threat' posed by the expansion of foreign banknotes issued by the neighbouring small German states.[21] The easy money policy by the unified German state from the 1870s would be better understood as an *ex post* acceptance of the private sector's lead than a state initiative.

### *The organiser of the industry and cartels*

The German investment banks were never passive financiers. They were entrepreneurs, perceiving new business opportunities and organising the industry for utilising those opportunities.[22] Although the investment banks were engaged in organising the iron and steel industry from the late 1850s, when newly established firms faced financial difficulties, the role of the banks in organising the industry became pronounced during the Great Depression in the 1870s.[23] It was during this period that it became a common practice for the banks to send their representatives to the management of enterprises.[24]

The cartelisation movement was the culmination of the organising role by the banks. It was a direct response to the Great Depression.[25] The banks were worried about the collapse of firms in which they had a stake. It was imperative for them to secure their investment in the industry. In this situation, cartelisation was an arrangement that could at least ensure the survival of firms although 'it did not guarantee high profits'.[26] Orderly management of the market suited more to the interests of the banks than cut-throat competition in a given market. The German banks even forced some recalcitrant producers like Krupp, which had not been quite dependent on them for financing, to join cartels using their financial power.[27]

Here again, the role of the German state cannot be disregarded. Whilst cartels in Britain did not get support from the state, those in Germany could enforce their contracts in the courts.[28] As the banks initiated cartelisation primarily for their own interests, so did the state support the cartelisation mainly for its social objectives like employment, social security, and so on.[29] Cartelisation was, however, beneficial to the competitiveness and expansion of the German iron and steel industry for the following reasons.

Above all, it provided a demand-side boost for sustaining large scale capital investment. Gerschenkron's schema is largely concerned with supply-side issues in the process of catching-up. That is, establishing bigger and bigger plants is a strategy of latecomers to compete with forerunners in a given market. However, capital-intensive investment is vulnerable to demand fluctuations. If demand is insecure, big plants are not always more efficient than small ones. Webb (1980: 324) therefore points out that '[o]nly if demand fluctuations were reduced would the most efficient, capital-intensive technologies always be the most attractive'.

A major function of the German cartels in the iron and steel industry was to preserve the domestic market for cartel members to mitigate market fluctuations. There is also the role of the state to consider. The German domestic market was protected by tariffs in addition to cartel arrangements. For, if imports of foreign products were freely allowed, cartels could not be sustained. In this respect, 'the tariff-cartel system' is a more appropriate term to describe the demand-side support given to the German iron and steel industry.[30]

In international competition, the strength of the tariff-cartel system lay in the possibility of export dumping. As Webb (1980: 324) argues, '[t]he tariff-cartel system could keep domestic prices high enough to cover fixed costs, while firms added to their profits by selling at marginal cost on the depressed world market'.[31] Fremdling (1991b: 133) therefore argues that 'it was the tariff-cartel system, supported by the state and enforced by the banks, that made Germany's exports more competitive on the world market than those of their British counterparts'.

It is still a matter of debate whether the German tariff-cartel system contributed to raising the efficiency of the German producers compared to their British counterparts.[32] But it is clearly true that the German tariff-cartel system positively contributed to the output expansion of the German iron and steel industry. From 1879 to 1913, the German steel output grew 10.4 per cent a year and pig-iron grew 6.3 per cent whereas British steel and pig-iron output grew 6.0 per cent and 1.5 per cent respectively for the same period.[33] By preserving the foreseeable market demand through the tariff-cartel system, the German producers could make a more vigorous capacity expansion than their British counterparts.[34]

And there were some grounds that the tariff-cartel system contributed to the efficiency of the German iron and steel industry, although a quantitative justification is not available to us. First, cartelisation enabled the selective exploitation of plants across industry during the period of depression. It should surely have had efficiency increasing effect to close down inefficient plants and to operate efficient plants more intensively.[35]

Second, cartelisation facilitated vertical integration of the German iron and steel industry. A cartel is in essence a *horizontal* coordination among firms for a limited period through contractual terms. In the German iron and steel industry, however, the cartelisation created incentives for individual firms to *vertically* integrate backwards and forwards. Backward integration was a way 'to avoid paying the syndicate prices for . . . inputs' and cartel members of intermediate products had 'an incentive to integrate forward into non-cartelized or weakly cartelized product lines, in order to utilize expanded capacity without violating the cartel contract for the intermediate good'.[36]

Vertical integration made possible several technical economies. Integration itself reduced the costs involved in material handling. It also enabled extensive mechanisation, such as the introduction of continuous conveyors, travelling cranes, and so on.[37] Moreover, vertical integration had a fuel-saving effect, 'by doing all processing "in one heat" without ever letting the metal cool down to where it required reheating'.[38] Actually, many of the major advances in steel technology from the late nineteenth to the early twentieth century were based on increased vertical integration.[39]

There were certainly some negative effects of the banks' close participation in the German iron and steel industry and the resulting cartelisation of the industry. The German banks were often accused of 'riskiness' or overconcentration of their investment in capital-intensive sectors. And the cartelisation movement underwent 'the era of power-seeking', not wholly led by economic considerations.[40]

In order to reach a rigorous assessment of the *net* effect of the banks' role in the industry, a broader study is surely required. In particular, those questions as to how and why the power of the investment banks was not misused, how they could act as entrepreneurs rather than as conservative financiers, and so on, should be answered. One critical factor which contributed to reducing possible negative effects of the banks' influence, would be the fierce competition among the banks. As Trebilcock (1981: 100) argues, '[b]etween the Great Banks, competition for business was as fierce as collaboration within joint ventures was close . . . [and] it is clear that sufficient competition survived to prevent the formation of "collusive" profit policies'.[41] Landes (1969: 350) therefore observes that, on the whole, the banks' influence 'was on the side of a more thorough utilization of resources and a more effective combination of the factors of production'.

## 6.2 BEGINNING OF THE MODERN IRON AND STEEL INDUSTRY IN JAPAN: A STATE-LED CATCHING-UP

On the level of the overall economy, the Japanese catching-up pattern was quite different from that of Russia in its sectoral composition. Whilst Russia, as well as Germany, concentrated its catching-up effort mainly on heavy industry, the Japanese catching-up was largely led by light industry, especially by the textile industry. In Japan, light industry was dominant in the initial period of industrialisation and remained significantly large even until the 1930s.[42] The growth rate of Japanese cotton fabric exports was 'the highest in the world' from 1890 to 1940 and Japan rose to the position of the largest cotten fabric exporter in the world in the early 1930s.[43] There were several reasons for these different industrial compositions between the two countries, which are beyond the scope of this book.[44]

As far as the iron and steel industry was concerned, however, the pattern of Japanese catching-up was very similar to that of Russia. This was mainly because functional imperatives in Gerschenkron's schema operated in Japan in the same way as in Russia. That is, the iron and steel industries of the two countries shared common factors such as 'extreme backwardness', 'urgency' for catching-up and technological trend towards increasing economies of scale.

### *Similarities with the Russian pattern*

Associated with its extreme backwardness, the role of the state was prominent in the deveopment of the Japanese iron and steel industry as it was in Russia. The modern iron and steel industry began in earnest only with the establishment of the state-owned iron works, Yawata, in 1901.[45] The Japanese state did not have any predetermined intention to initiate its industrialisation by the public sector.[46] Rather, it was the general rule for the Japanese state to entrust the development of heavy industry to the *zaibatsu*. The Japanese private sector had also grown rapidly during the late nineteenth century, as reflected in the privatisation of government assets in the 1880s. As far as the modern iron and steel industry was concerned, however, the required scale of finance was still too big and technological barriers were too formidable for the private sector to surmount in a short time. For example, one powerful *zaibatsu*, Sumitomo, had to abandon its attempt at establishing a modern iron works in 1894 because of technical deficiencies.[47] After the end of the Sino–Japanese War in 1895, the Japanese government asked the *zaibatsu* to establish modern iron and steel mills without success.[48] Even after the establishment of Yawata, the private sector was still hesitant to be fully committed to the iron and steel industry, engaging only in peripheral operations because even Yawata continued to make losses for its first nine years, surviving largely on the state demand.[49]

Of course, Japan could have taken a more gradual route for establishing its modern iron and steel industry by waiting for its private sector to grow enough to undertake the risks involved. But the driving force of Japanese industrialisation was the military imperative, as in Russia. After having its border opened up by force and having to witness China being divided and ruled by Western imperial countries, the Japanese state regarded its most urgent task as building military power through heavy industrialisation.[50] In establishing the modern iron and steel industry, a government document clearly emphasises the urgency for heavy industrialisation: 'If we wait upon private interests . . . we will not be able to anticipate a time when we will have a satisfactory large-scale iron industry in this country.'[51]

If the question for Russia had to do with whether it could maintain its position in the European power politics, for Japan it was whether the country could survive in the face of the imperial aggression from the West. In this sense, the forced opening of the Japanese border by Perry in 1854 played the same role as the ignominious retreat of Russia in the Crimean

War in 1857 and in the Berlin Congress in 1878. It made it clear to the Japanese leaders that catching-up with its forerunners through industrialisation was a task that could not be postponed any longer, as was also the case in Russia. Moreover, its private sector was then not developed enough to initiate the catching-up process at the required speed. Thus, '[i]t was the State that conceived modernization as a goal and industrialization as a means, that gave birth to the new economy in haste and pushed it unrelentingly as an ambitious mother', as Landes (1965: 182) points out.

In its bid to achieve a rapid catching-up in the iron and steel industry, Japan attempted a Gerschenkronian investment. That is, Yawata was established as an integrated iron and steel works with the most advanced equipment from the beginning. The first head of Yawata, Takato Oshima, 'tried to imitate the German model as precisely as possible'.[52] And '[t]he scale of plant was . . . almost on a par with the developed countries'.[53] From the beginning, there was a clear aim of rapid technological catching-up, as there was in the Russian case.

### *Deviations from the Russian pattern*

However, the Japanese catching-up speed was much slower than that of Russia. Yawata iron works 'had to stop operation for two years' just after its opening in 1901 because of technical problems.[54] The pace of technological absorption was very slow.[55] The output also increased very slowly. Even by 1913, the Japanese producers could supply only 48 per cent of the pig-iron and 34 per cent of the steel needed for domestic consumption.[56] Yawata iron works could not achieve the technological level of its forerunners in the West for almost two decades from its establishment. Its competitiveness before the 1920s rested mainly on extremely low wage costs and moderate protective tariff of around 15 per cent.[57]

This relatively slower pace of catching-up by the Japanese iron and steel industry can be attributed to the limited extent of *substitutes* that could be employed by the Japanese state. First, Japan was very restrictive in attracting foreign capital since its policy-makers were preoccupied with the question of building an *independent nation* invulnerable to the imperialist aggression from the West. When the issue of attracting foreign loans to mitigate fiscal difficulty was discussed in 1880, '[t]he Emperor's decision was against a foreign loan as too dangerous'.[58] The official policy of the government was 'virtually to exclude foreign business investment, except in trading activity'.[59]

Moreover, as Rosovsky (1961: 120) points out, '[f]oreign borrowing was limited by the uncertain credit worthiness of a "new" nation'. Even if Japan had wanted to raise foreign capital, it would have been very difficult. The consequence was that the contribution of foreign capital was negligible in Japanese industrialisation. 'The initial Japanese advances – in textiles, metallurgy, even railways – were accomplished with domestic savings almost exclusively.'[60] This situation stood in stark contrast to the case of Russian industrialisation, which was heavily dependent on foreign capital. The French government was the paymaster for the Russian railway construction mainly for its strategic military purpose.[61] Roughly one-third of capital in Russia was supplied from outside in 1890 and half the joint stock capital was held abroad by 1900.[62]

Second, Japan was also very restrictive in foreign borrowing of technologies. In Russia, '[a]s instructors in western "best practice" methods, foreign entrepreneurs made *decisive* contributions to Russian technological performance'.[63] But, in Japan, 'outsiders were brought in for specific purposes – . . . a few experts [were] "rented" . . . today here, tomorrow there – kept only as long as necessary, and sent away'.[64] This was because Japan feared foreign dominance. Moreover, the employment of foreign engineers was simply too expensive and there were cultural differences which made it difficult for foreign engineers to stay in Japan.[65]

Third, the Japanese state was more bound by fiscal difficulty than the Russian state. The Russian industrial take-off in the 1890s was commenced on the basis of the ruthless budget balancing measures undertaken by Vyshnegradski (1887–1892), Finance Minister prior to Witte.[66] By contrast, the Japanese state faced a severe deficit problem all through the latter half of the nineteenth century. Within a decade of the Meiji Restoration in 1867, the Japanese government issued large amounts of cash and government bonds to cope with 'the *samurai* problem', carried out one foreign expedition to Formosa, and experienced one major civil war. By 1880, the state budget was 'virtually bankrupt'.[67] The main purpose behind the privatisation of state enterprises in the 1880s was to alleviate this fiscal problem.[68]

The Japanese iron and steel industry achieved the level of technical efficiency comparable to that of advanced countries only in the 1920s with an extensive rationalisation programme. This time, however, the state was no longer sole organiser of the industry. Just as the Russian economy entered a new 'stage of backwardness' with emerging investment banks in the early twentieth century,[69] the Japanese iron and steel industry moved on to a new stage with the growing capacity of the *zaibatsu*. The task of the rationalisation was shared between the state and the *zaibatsu*.[70] And some *zaibatsu* iron works began to appear more aggressive in their investment than the state iron works.[71]

## 6.3 CONCLUDING OBSERVATIONS

The German catching-up in the iron and steel industry was indeed a typical case of bank-led growth. As Gerschenkron emphasises, the investment banks provided the German industry with 'an organisational direction' which contributed to overcoming disadvantages and exploiting advantages of the latecomer in the latter half of the nineteenth century. In fact, what the investment banks actually did was to devise and implement a 'semi-industrial policy'. Their broad industrial interests enhanced their vantage-point in devising an industry-wide policy. And their dominant power over the industry was indispensable in implementing their policies.

This role assumed by the investment banks certainly had to do with the moderate backwardness of Germany during the period. The German industry still had to be more organised in order to compete with its forerunner, Britain. But its private sector had already developed and become more complex, and state interventions had begun to show 'stultifying effects' rather than inducing effects for industrial growth. Here, the investment banks were an alternative institution to organise the industry. Although the role of the state was important, as we saw in its tariff policy, it was largely confined to supportive actions rather than a direct promotion. As Trebilcock (1981: 99) argues, '[t]he implicit ability to charter inter-industry connexions, to promote structural change, or to engineer cartelization resided more with the Great Banks than it did with the imperial state'.

The Japanese catching-up in the iron and steel industry during the early twentieth century also broadly fits Gerschenkron's model of extreme backwardness, very similar to the case of Russia. But it shows a case of a *bound* state: the Japanese state was politically, economically and geographically bound in introducing foreign capital and technologies. Thus, the speed of catching-up in the Japanese iron and steel industry was slower than that of Russia. A similar case of a bound state and subsequent gradual route for catching-up can be found in the case of the Korean semiconductor industry, as shall be discussed later (section 8.4).

On the level of the overall economy, the Japanese relatively slower growth in the iron and steel industry, and heavy industry in general, as compared to the Russian case, seemed to have been compensated by its faster advance in light industry, as discussed before. Since our investigation was confined only to the iron and steel industry, however, we did not provide satisfactory explanations about differing patterns of catching-up at the economy level and their efficiency implications.

# 7

# USA–JAPAN–KOREA COMPARISON IN THE IRON AND STEEL INDUSTRY, 1950–1990

The first half of the twentieth century was a period of the USA 'forging ahead'. The USA output of crude steel already accounted for 36.8 per cent of global output in 1900. That share was increased to 42.9 per cent in 1940 and reached more than 50 per cent in 1947.[1] During the period, major technological and organisational breakthroughs were also originated in the USA. USA steel producers perfected the open-hearth process, increased the scale of operation dramatically, and introduced 'the hard driving' method.[2]

What should be noted, however, is that the characteristics of the American dominance during the period was quite different from that of the British dominance during the nineteenth century. The former was mainly sustained by the vast and rapidly growing internal market whereas the latter was highly oriented towards export. Until the outbreak of the First World War, the USA domestic market was heavily protected by tariff.[3] During the war periods and the interwar period, its European competitors, preoccupied with the war efforts and post-war reconstruction, did not have spare strength to challenge the American market.

Moreover, competition within the USA domestic market was not so fierce once the dominant leadership of US Steel was established in 1901 through a series of mergers.[4] US Steel was thereafter preoccupied more with the political risk control in the 'Progressive Era' (generally 1900 through 1920), than with innovation.[5] Thus, although the output level and labour productivity of the USA iron and steel industry remained the highest in the world all through the first half of the twentieth century, its pressing reorganisation in the 1930s was postponed.[6]

The leadership challenge from Japan and later from Korea during the post-war period was basically that of a globalised production system against that of a domestically oriented production system. The technological trend to the iron and steel industry was still towards increasing economies of scale. The ingenious aspect of the Japanese catching-up strategy was to organise its production on the global scale and thus to utilise the technological trend to its full extent. Korea imitated the Japanese catching-up strategy just as Russia had imitated that of Germany over seventy years before. Due to the different degrees of backwardness, however, there were differences in the institutional pattern of catching-up between Japan and Korea similar to those between Germany and Russia. Gerschenkron's schema broadly fits the catching-up process of Japan and Korea in the iron and steel industry during the post-war period. An elaboration of the reasons for this recurrence of Gerschenkron's pattern, as well as some deviations, is the purpose of this chapter.

## 7.1 JAPANESE OVERTAKING OF THE USA: OLIGOPOLISTIC COMPETITION AND MITI GUIDANCE

With the end of the Second World War, USA leadership in the iron and steel industry appeared to become more robust. The industry's greatly increased production capacity during the war period remained untouched while major producers in Europe and Japan struggled to recover from the war damages. The USA iron and steel industry also undertook 'its largest capacity expansion in history' in the 1950s, responding to the increasing demand in the world economy.[7] Nevertheless, leadership of the industry shifted to Japan in a couple of decades. Why?

### *Functional aspects of the Japanese overtaking*

As mentioned before (section 4.2), the Japanese overtaking of the USA was closely associated with the development of new technologies, notably the basic oxygen furnace (BOF) and continuous casting (CC), and their effective utilisation. By 1962, 'all of the major integrated Japanese steel-makers were operating BOFs', whereas it took 'nearly another decade' for USA steel-makers to complete their transition to the BOF technology.[8] CC spread at the extent of 70 per cent in Japan by 1981 but the rates were only 21 per cent in the USA and 45 per cent in the EC by the same time.[9]

There were certainly some elements of latecomers' advantages in the speed of Japanese adoption of the new technologies. The USA producers, partly due to their dominant position as the world leader, did not have a pressing need to experiment with the unproven BOF technology in their capacity expansion in the 1950s. Then, BOF had not been tested in large-scale operations and, above all, the quality of its product was suspected because the steel-making process was too fast.[10] The USA government procurement policy during the Korean War (1950–1953) added an incentive for the USA producers to stick with the proven technology, i.e., the open-hearth furnace (OHF).[11]

By contrast, the Japanese producers were more willing to take the risks involved in the unproven technologies. First of all, BOF was extremely economical.[12] For Japanese producers, who were financially more constrained as latecomers, this economic advantage was more attractive than for USA producers. BOF technology was also viewed as a way of overcoming their dependence on scrap, stable acquisition of which became increasingly difficult in the 1950s.[13] Moreover, the timing of capacity expansion seemed more favourable to the Japanese producers. Their major capacity expansion began in 1955 with the Second Rationalisation Programme (1955–1960), rather later than that of their US counterparts. The Japanese producers had less burdens from previous investments on the old technology and there had been some improvements in BOF.[14]

The adoption of CC was closely connected with that of BOF. US and European producers were cautious in adopting CC because it had many technical problems in application and maintaining quality, on the one hand, and because the cost incurred in building the CC plant was considerable, on the other. For Japanese producers, however, CC was complementary to their BOFs because 'it enabled makers to cast the large masses of steel produced by the BOF faster and more efficiently than in conventional slab mills'.[15] Japanese steel-makers countered the quality problem arising from the fast steel-making process in BOF and CC by developing computerised process control earlier than any country.[16]

It seems to us, however, that the most distinctive feature of the Japanese catching-up in the iron and steel industry during the post-war period was its global production system. All of the major steel producing countries had previously based their production systems mainly on domestic raw materials. Japan became the first country in the history of the iron and steel industry to rise to the position of world leader without benefiting from raw materials of its own or adjacent areas.[17] Japan imported raw materials from all over the world, established plants big enough to exploit economies of scale on the global level and exported its products to the world market.

What is important to note is that the basic model for this global production system was already formulated and experimented *before* the introduction of BOF and CC. Nishiyama, the head of Kawasaki Steel, proposed this model as early as in 1950 and established the Chiba works with OHFs, not with BOFs. Nishiyama's model was in essence a Gerschenkronian one. He proposed that Chiba works were 'to be equipped with the world's most advanced technologies in the best factory layout' in order to 'compete internationally'.[18] This meant the establishment of 'bigger and bigger' plants since the technological trend in the iron and steel industry was still towards increasing economies of scale. And this Gerschenkronian investment strategy was combined with the Japanese peculiarity, i.e., its lack of domestic raw materials. Nishiyama located the new plants 'on the sea-coast with deep-water port', aiming at reducing the costs in importing raw materials and shipping intermediate and final products.[19]

In this respect, BOF and CC are better understood as corroborating factors for the Japanese catching-up model, rather than decisive factors in shaping it. On the basis of Nishiyama's experiment, Japanese producers adopted the global production system and were quick to utilise newly developed BOF and CC when they embarked on capacity expansion from 1955. Japanese producers were also very active in developing foreign raw material, especially in Australia.[20] They were innovative in reducing the transportation costs as well by developing new types of energy-saving ships and organising a special cargo company, the Japan Iron Ore Transporting Company.[21] And it was imperative to actively engage in export activities in order to pay for imported raw materials, on the one hand, and to realise the dramatically increased production capacity, on the other.

It was mainly the result of the adoption of the global production system that Japan could have the eight biggest steel mills in the capitalist world by 1977.[22] The dramatically increased economies of scale on the global level were a crucial factor in the Japanese international competitiveness.[23] Such a concentration of huge production capacity in one country was only possible with the exploitation of the global market. The Japanese dependence on the export market was thus very high, running at around one-third of its steel output in the post-war period.[24]

### *The* keiretsu *and oligopolistic competition*

The institutional pattern of the Japanese catching-up in the iron and steel industry also broadly fits Gerschenkron's moderate backwardness. As Japan had already undergone a substantial period of industrialisation during the first half of the twentieth century, its catching-up during the post-war period was based more on private initiative compared to that about half a century before. This private sector initiative is more evident if Japan is compared to its follower, Korea (section 7.2).

Although the Japanese state played an important role in this catching-up process, as will be discussed below, its initial attitude towards the prospect of the iron and steel industry was rather conservative. It remained basically a follower and coordinator of the private sector initiatives, at least as far as the iron and steel industry was concerned.[25]

For instance, the Japanese state at first opposed Nishiyama's plan, regarding it 'as an impossible dream'.[26] MITI defined the iron and steel industry as an 'inappropriate exporting industry'.[27] Its initial rationalisation effort was therefore focused on the later stage of steel-making such as rolling processes, and hot and cold strip mills, rather than on establishing integrated mills.[28] MITI's position changed only with the unexpected steel export boom in 1955. MITI adopted 'the expansion and construction of an integrated production system' as one of its policy objectives only in the Second Rationalisation Programme (1955–1960), five years after Nishiyama's proposal.[29]

This conservative attitude of the Japanese state can be attributed to several factors. The Japanese state was under the control of the Supreme

Command for the Allied Powers (SCAP) until 1952 and it had to follow the Dodge Line, which strictly required a balanced budget. After the end of the Occupation, it also took time to overcome the internal disagreements within the state.[30] Viewed in a comparative perspective, however, the moderate backwardness of the Japanese economy was pronounced.

As mentioned before (section 6.2), the *zaibatsu* had already actively participated in the rationalisation of the iron and steel industry even in the 1920s. The capability of the Japanese private sector was further enhanced with the heavy industrialisation in the 1930s and during the war period.[31] Most iron and steel mills in Japan remained intact after the war although production facilities in Manchuria were cut off from Japan. Experienced managers and engineers, and well-trained workers were still abundant, although some top executives were purged.[32]

This situation resembled that of the German economy in the latter half of the nineteenth century (section 6.1). Japan had already reached the technological and organisational level to exploit new technologies without any significant delay. Japan's problem just after the war had to do more with the lack of financial resources and the constraints imposed by the international situation, than with its own technological and organisational capability.[33]

In this context, it seems important to pay attention to the change of the competitive structure *within* the Japanese iron and steel industry after the Second World War. Until then, the private non-integrated steel firms like Kawasaki, Sumitomo and Kobe had not been in direct competition with Japan Steel (Nippon Seitetsu).[34] Japan Steel provided 'cheaper and better iron and semi-finished steel under the strict control of the government'.[35] But SCAP divided Japan Steel and created two private firms, Yawata and Fuji, as one of its attempts at debilitating Japanese military capability. These two newly born giants now became direct competitors against previously non-integrated steel-makers.[36]

Nishiyama's initiative was basically a response to this newly emerging competitive structure in the domestic market. It was based on economic rationale for Kawasaki to integrate backwards to pig-iron production and to establish 'bigger and bigger' plants in order to compete with Yawata and Fuji. For the sake of its own interest, Kawasaki Steel started the construction of Chiba works in 1951 'without MITI's approval'.[37] Other private non-integrated steel-makers such as Sumitomo and Kobe followed suit because they faced the same competitive pressure as Kawasaki. The consequence was the emergence of *oligopolistic competition*. A small number of big integrated steel producers began competing fiercely with each other.

The Japanese peculiarity was that this oligopolistic structure was combined with the (horizontal) *keiretsu* system, whereas it was combined with the investment banks' domination in Germany about a century before.[38] The principal role of the *keiretsu* lay in the provision of capital, as did that of the German investment banks. The *keiretsu* facilitated the high-speed economic growth by enabling 'overborrowing'.[39] Mainly for the purpose of finance, previous *zaibatsu* ties were strengthened and new ties were established. For instance, Kawasaki Steel established a close alliance with Daiichi Bank in order to finance its ambitious investment plan. Sumitomo Steel was supported by its *keiretsu* member bank, i.e., Sumitomo Bank. Yawata and Fuji, though they were classified as independent from any *keiretsu* grouping, strengthened their existing links with the Mitsui and Mitsubishi *keiretsu*.[40] NKK's joining Fuyo group was also driven by financial necessities because its affiliated *zaibatsu*, Asano *zaibatsu* lacked financial institutions.[41]

However, there was one significant difference between *keiretsu* and the German investment banks: that of coordinating power. In Germany, the investment banks were *de facto* the centre of German capitalism. They not only provided the industry with capital but also enforced industry-wide reorganisation with their multiple holdings across competing firms. By contrast, competing large firms were mostly backed by competing banks in Japan.[42] So competition was not restrained, rather intensified, by the banks' support. The coordination mechanism operated only *within* the (horizontal) *keiretsu* system. There was no effective inter-*keiretsu* coordination mechanism within the Japanese private sector. This would be why many Japanese business historians speak of 'excessive competition'.[43]

Moreover, as Yonekura (1991: 227) argues, the situation of 'overborrowing' and investment on a greatly increased scale made it more difficult for individual producers to operate at a low utilisation rate. For they had to pay periodic interest on borrowed capital and their break-even points had risen significantly due to large-scale investment. This drove them to compete fiercely for market share and for exploring new markets, and made it difficult for them to agree on production restraints.

Under these circumstances, the Japanese state, especially MITI, was a substituting institution for industry-wide coordination. It would require further research to establish whether MITI was as effective an institution as the German banks in enforcing cartels.[44] It seems that MITI's intervention was not so effective during the period of vigorous expansion, but that it began working once the problem of overcapacity became acute, similar to the German case, where cartels were basically 'children of financial storms' during the Depression.[45]

MITI has continually attempted to control investment and sales from the beginning of the 1950s, as was reflected in the designated-price system (1950), the Steel Supply and Demand Stabilisation Law (1956), the open sales system (1958), and so on. But these were all abandoned upon confrontation with 'unexpected' booms or oppositions from the industry. MITI's outright intervention changed its appearance with voluntary self-regulation (1962); even this was challenged by the 'Sumitomo Rebellion' in

1965. MITI finally found a solution for the problem of excessive competition only by destroying the structure of oligopolistic competition itself. It supported the merger between Yawata and Fuji, producing a price leader, Nippon Steel, in 1968.[46] At this time, even private firms supported the merger because they acutely needed a solution for cut-throat competition with overcapacity.[47]

### *The developmental state*

The Japanese state certainly played the 'negative' role as the German state had done in the previous century. However, it intervened in its industries more directly than its German counterpart, as we saw above, in its continual attempts at cartelisation of the iron and steel industry. We have pointed out that this occurred because Japan lacked the centre of coordination within the private sector. But it was also because the Japanese state's power to intervene in the economy was increased significantly.

The situation of the Japanese state after the defeat in the Second World War was historically unique. It was not allowed to divert its ability to military affairs. Its role in international politics was severely restricted. The only remaining option for the Japanese state to enhance its national prestige was through economic development.[48] Moreover, its relative autonomy over domestic interest groups was significantly increased, '[w]ith the *zaibatsu* weakened, the military smashed, and the landlords dispossessed, but with the bureaucracy untouched'.[49] This increased relative autonomy of the state was corroborated by the one-party domination of the domestic politics. Hence the emergence of *'the developmental state'*. Economic growth was regarded as the prime objective of the state and the state was equipped with power to pursue that objective.[50]

One important role of the Japanese developmental state was its macro-level stimuli to industry. Its ambitious growth goal itself was an inducement for active investments at the level of industry.[51] 'Easy-money policy' was a crucial policy measure in facilitating macro-economic expansion. As the German Reichsbank's generous re-discount policy allowed the German investment banks to lend to the hilt, the Japanese state's generous monetary policy provided its *keiretsu* system with an environment in which it could keep overborrowing. The Japanese overborrowing system was actually twofold, as in Germany: the banking system as a whole was 'overlent' through the central bank's lending to the banks, on the one hand, and the industrial sector was overlent through the *keiretsu* system, on the other.[52]

But there was a pronounced peculiarity of the Japanese developmental state: the comprehensive employment of industrial policy in achieving its macro-economic goals.[53] We have pointed out that (semi-) industrial policy was formulated and implemented by the investment banks in Germany (section 6.1). The German state did not have any effective means to affect the banks' allocation of their credits. The Japanese state, however, had controlling power over banks' interest rates, dividend rates, introduction of foreign capital, office facilities, and so on.[54] With this power over banks, the Japanese state could influence the banks' decisions on credit allocation.

The Japanese government used the Japan Development Bank (JDB), a government-owned specialised bank, as a major tool for allocating credits for its industrial policy purposes. The quantitative share of JDB loan was not so significantly high except for a few years at its beginning.[55] But it had an important 'pump-priming' effect because the state asked commercial banks to extend 'cooperative loans' to the designated industries when JDB provided loans to these industries.[56] So Ueno (1980: 403) summarises the Japanese credit allocation system as follows:

> broadly speaking, the total supply of funds in Japan was controlled by the Bank of Japan, the level and structure of interest rates were artificially regulated by the Ministry of Finance, and private funds were allocated, under the guidance of public financial institutions, by city banks which competed for market shares.[57]

As one of the designated heavy industries, the iron and steel industry benefited from the state's influence on credit allocation.[58]

The Japanese state also maintained a tight protection system as a part of its industrial policy. It provided its heavy industry with almost 'absolute protection' by various means.[59] The iron and steel industry was no exception. The import of foreign iron and steel was 'generally prohibitive' until the mid-1960s.[60] The protective measures were important, especially in the 1950s when the Japanese iron and steel industry had not yet gained competitiveness against its American counterpart.[61] Even with liberalisation measures in the 1960s, 'foreigners were prohibited from supplying the precise kind of steel needed by the automobile industry'.[62] The post-war IMF–GATT system was applied only nominally to Japan as it was to Western Europe, though explorations of the reasons for this would be beyond the scope of this book.[63]

The Japanese state also directly intervened in the process of technology transfer. Its intervention in the introduction and diffusion of BOF was a classic example of technology policy.[64] The Japanese state was more directly involved in organising its industry as well. MITI's role in '*keiretsi*fication' of the industry was important here.[65] And, as we mentioned above, MITI has kept attempting cartelisation of the iron and steel industry.

Further detailed research would be required to assess how much the Japanese state's direct intervention in the iron and steel industry contributed to its growth as compared to its inaction. Since its private sector had already developed enough to initiate its ambitious catching-up, the Japanese state's intervention might not have had as great an inducing effect as that in the beginning of the twentieth century. But, as emphasised above, the Japanese state's industrial policy was substituted for the absence of coordination within its private sector. And its emphasis on heavy industry certainly benefited the development of the iron and steel industry.[66]

## 7.2 KOREAN CATCHING-UP: POSCO AND 'KOREA INC.'

The Korean catching-up in the iron and steel industry during the post-war period was almost solely undertaken by one state enterprise, the Pohang Iron and Steel Company (POSCO).[67] Before POSCO came into operation in 1973, domestic output of steel was very small and was produced entirely from non-integrated plants.[68] In less than two decades, however, POSCO grew into one of the most cost-efficient steel producers in the world. By 1990, it was ranked as the third-largest producer in the capitalist world. Mainly due to the production increase in POSCO, Korea had become the eighth-largest steel producing country in 1989, surpassing such industrialised countries as France, the UK, Canada and Spain.[69] POSCO even exported its technologies to the USA in 1986.[70]

POSCO faithfully followed the Japanese model, which was basically a Gerschenkronian one. Its first integrated mill was constructed with instruction and technical assistance from Nippon Steel, then the most advanced steel producer in the world. The principle of establishing 'bigger and bigger' plants was also applied. The earlier stage of POSCO's establishment can be described as a struggle for attaining a bigger plant.[71] The latest technologies were also introduced. The first blast furnace was, of course, based on BOF. CC and computerisation were extensively adopted.[72] Currently, POSCO boasts the most modern facilities and the best world levels of steel production.[73] Moreover, its mills were built on the sea-coast near deep-water ports. Its production system was organised on a global scale, importing most raw materials from abroad and giving strategic importance to export activities.

### *The developmental state and heavy and chemical industrialisation*

Although the Korean catching-up strategy in the iron and steel industry was almost a duplication of the Japanese model in its functional aspects, its institutional underpinning was different from the Japanese, reflecting its relative backwardness. That is, the Korean catching-up was initiated by the state, as POSCO was established as a state enterprise. The Korean private sector simply could not contemplate this large-scale venture due to its 'extreme' backwardness, although industrial growth was by that time well under way led by the exports of light industrial goods like textiles, plywood, wigs, and so on, in the 1960s.[74]

The most critical difficulty was a financial one. POSCO's establishment was Korea's 'largest single investment to date'.[75] Its initial foreign capital requirement was $138.7 million, almost 70 per cent of total capital requirement of $200 million. This amounted to more than one-fifth of Korea's total exports in 1967 when POSCO's construction began.[76] To our best knowledge, there had been no attempt from the private sector to establish integrated mills in the 1960s. Instead, there had been several attempts, at least four, by the government from 1958, all of which collapsed due to the failure in raising foreign capital.[77]

Even POSCO's establishment almost failed because IBRD refused to support it. POSCO was able to raise its initial capital only through changing the usage of a large portion of reparations money paid by the Japanese government which had already been earmarked for the adjustment of the agricultural sector, and through obtaining loans from the Japan Export–Import Bank, neither of which could be secured without the state's absolute support.[78] The Korean government even conceded to the Japanese government on the mutual taxation agreement in order to secure Japan's cooperation in the establishment of POSCO.[79]

Of course, the Korean state might have postponed this large-scale venture until its private sector could grow enough to undertake it. In fact, this was the backbone of the IBRD recommendation when they refused to support the establishment of POSCO. But the actual course taken by the Korean state was to precipitate itself into a vigorous drive towards heavy industrialisation. The establishment of POSCO and the construction of a chemical complex in the late 1960s marked the beginning of its ambitious heavy and chemical industrialisation (henceforth HCI) programme during the 1970s.

In this 'drive' for HCI, there was certainly an element of the 'developmental state', as there was in the Japanese state during the post-war period.[80] In Korea, during the period of President Park's regime (1961–1979), other social classes which would stand against the state were kept out of political influence. Already the landlord class had been smashed with sweeping land reforms before and during the Korean War. Political elites penetrated by 'superannuated landlords' of the previous Rhee regime were removed in the military coup of 1961.[81] Instead, the government bureaucracy was modernised and strengthened.[82] There was no influential organised working class even up until the mid-1980s. And, unlike the previous regime, the new regime placed its top priority on industrial development.[83] Moreover, the industrial development became the only source of legitimisation, since the regime considerably restricted the democratic process in the name of economic growth.[84]

The Korean HCI in the 1970s was largely a necessary measure to further economic growth, as was the Japanese HCI in the 1950s and 1960s. As MITI saw it as 'the only way to break out of Japan's inevitable balance of payments constraints',[85] so the Korean state regarded it as inevitable to counter its chronic BOP problem.[86] The BOP problem was worsening throughout the 1960s despite the impressive export growth based on light industry because the export growth was more than compensated for by the growth of imports of raw materials, parts, equipments and technologies from abroad.[87] Moreover, the Korean economy faced the increasing need for creating new markets and developing new products due to the rising protectionism in the USA as well as due to the increasing catching-up pressure from its followers in South East Asian countries.[88] The competitive edge of the Korean light industries, primarily based on low wages, was rapidly being eroded.

Unlike the case of the Japanese developmental state, however, there were non-economic imperatives in the Korean state's drive for the HCI. The lukewarm US response to the North's partial military provocations in 1968,[89] and President Nixon's commitment to the gradual withdrawal of US forces in the South were facilitating factors for the HCI as a way to build its own miltary industry. The Korean state in fact extracted concessions regarding assistance in building its military industry from the USA as a compensation for the withdrawal of the troops.[90] Moreover, the HCI was regarded as a necessary step for the Korean state to achieve its goal of 'self-reliance', which meant not only being more independent from foreign resources and technologies but also being more independent from the US military presence and the resulting interference of the USA in domestic affairs.[91]

It would be difficult to assess precisely which factors, economic or non-economic, were more decisive in shaping the HCI programme.[92] But the non-economic aspect of the Korean HCI programme was very similar to the Japanese state's fostering of heavy industry in its modernisation around the turn of the century. Light industry might have been the engine of economic growth during the earlier period of modernisation. But the heavy industrialisation was indispensable in building military power, on which the state's support was concentrated, as discussed before (section 6.2).

The Korean HCI programme resembled that of the earlier Japanese state in another respect. That is, due to 'extreme' backwardness of its economy, the state initiated and dictated the course of the HCI. Although the *chaebols*, the Korean counterpart of the Japanese *zaibatsu*, had existed from the 1950s, they were not powerful enough to launch large scale investments by themselves. In particular, their capital mobilisation was critically dependent on the state. And the Korean state even forced some reluctant *chaebols* to undertake new heavy industries.[93] The growth of *chaebols* in Korea was mainly the result of the state's initiative in the HCI as that of *zaibatsu* in Japan, not vice versa.

### *Peculiarities of POSCO*

Although the HCI was basically initiated by the state, it was a general rule for the Korean state to entrust the actual management of business largely to *chaebols*. The establishment of POSCO as a state enterprise appears an exceptional case in the broader process of the HCI programme. One reason had surely to do with the incompetence of the private sector in raising large amounts of initial capital for construction of modern integrated steel mills, as discussed above. Yawata was also established as a state enterprise mainly because of the incompetence of the Japanese private sector in raising initial capital and absorbing advanced technologies (section 6.2).

There was also a problem of 'natural monopoly' in POSCO's case. Unlike the other HC industries, the scale of the modern integrated mill was too large to allow a competitive structure in the small domestic market of Korea. Monopoly seemed inevitable, especially in the early 1970s. In this situation, the state might 'not want to give a monopoly to any one chaebol over an input that impinged on both the country's security and the competitiveness of major downstream industries', as Amsden (1989: 295) points out.

From a comparative perspective, however, it seems important to pay attention to one distinctive feature of the Korean iron and steel industry: the relation between POSCO's monopolistic position in the domestic market and the overall export promoting strategy of the Korean state.

In the case of the German iron and steel industry in the late nineteenth century, the coordination by the investment banks was rather *ex post*. They rarely controlled the capacity expansion *ex ante* (section 6.1). Moreover, it seems that the German industry as a whole had not been critically dependent on export activities, and the other heavy industries using steel products had already developed to a considerable extent. Although the introduction of high import tariffs on iron and steel in 1879 was mainly due to the political alliance of 'iron and rye',[94] this would have been difficult to sustain if the German steel-consuming industries were in a very weak competitive position or if their export activities were crucially important in the overall German economy.

In the case of the Japanese iron and steel industry during the post-war period, the state's controlling mechanism worked only *ex post*, as in the German case (section 7.1). And no particular role was assigned to its iron and steel industry to support other industries. In the case of the earlier period of modern Japan, Yawata retained a monopolistic position as a state enterprise. But it was supported by other industries for more than a decade, not vice versa (section 6.2).

A peculiar feature of the Korean iron and steel industry was that the monopolistic position of POSCO was combined with its assignment to support other heavy industries within the overall policy framework of export promotion. As Juhn (1991: 274) puts it, '[t]he company's initial objective was to set itself up as a catalyst for the industrialization of Korea by supplying an increasing volume of low-cost steel to steel consuming industries'. The Korean state intended to use the company as an instrument of its industrial policy. This would be why the establishment of POSCO was a bit earlier than the full-fledged drive for the HCI in the 1970s.[95] The Korean state also developed other industries related to basic materials like oil, coal, gas, fertiliser and electricity prior to the HCI programme. These industries were, like POSCO, owned by the state, and their products were supplied to designated strategic industries at subsidised prices.[96]

On the other hand, related to the overall export promoting policy of the Korean state, POSCO received a porous protection compared to the other HC industries.[97] There was nominal 25 per cent duty on the imports of steel products. But this did not provide POSCO with effective protection because, as one of the export promotion schemes, the Korean steel users could receive duty drawbacks if they re-exported the imported steel in a more processed form. This was a major reason why POSCO had to sell its products to local exporters at prices lower than import prices from the very beginning.[98] POSCO was sheltered from foreign competition only in the products supplied to end users who made products for domestic consumption.

Moreover, POSCO set itself an export target of 30 per cent for the following reasons:

1 to ensure long runs and full-capacity utilisation for all types of steel;
2 to earn hard currency to repay its foreign debts;
3 to take advantage of the government's export subsidies;
4 to drive a stake in the international market in anticipation of future capacity expansion.

In fact, its export ratio to total sale sometimes, for instance, in 1974, 1975 and 1982, exceeded even 40 per cent. Thus, the actual protection POSCO received from the tariff has been very low.[99]

The combined result was that it was difficult for POSCO to behave as a monopolist. Although POSCO set the prices for domestic users slightly higher than those for local exporters or foreign buyers in its earlier operations in the 1970s, there is little evidence that POSCO carried out export-dumping at the cost of domestic users.[100] Moreover, it kept supplying its steel products to domestic users at prices lower than foreign suppliers offered, although it could sell its products in the international market at better prices.[101] In this respect, POSCO's monopolistic position was not destined to extract monopoly profits, but to contribute to the overall economic development, symbolising the overall direction of the Korean state's intervention in the economy.[102]

A remarkable accomplishment of POSCO is that, despite the somewhat patchy protection it was given and despite certain sacrifices for the state's industrial policy purposes, it grew into a competitive enterprise in a very short time. It has remained highly profitable from the first year of its operation in 1973. With the continuous accumulation of handsome profits, the capital needed to build its second steel complex at Kwangyang, construction of which began in 1981, was raised solely through its own internal resources and its own international credit.[103] And its volume of steel export to Japan began to exceed that of Korea's import from Japan in 1981.[104]

As we emphasised above, the Gerschenkronian type of investment strategy, that is, the establishment of 'bigger and bigger' plants with the latest technologies, should have been a critical factor in POSCO's earlier attainment of international competitiveness and profitability. Various forms of government support also have been important.[105]

Viewed from the point of industrial organisation, the ongoing high-capacity utilisation due to POSCO's monopolistic position, needs to be brought to our attention. In the case of Japanese producers during the post-war period, their structure of oligopolistic competition made it difficult to adjust their capacity expansion collectively. But POSCO could adjust its capacity expansion based on calculation of future demand increase and maintain almost full-capacity utilisation.[106] If Korea had a competitive structure of industry, this level of high-capacity utilisation would have been impossible to attain.

There were also very important firm-level factors for the earlier attainment of POSCO's competitiveness, which cannot be included in our comparative schema directed at the industry-level issues, but were no less important than the factors mentioned above. For instance, POSCO could save on construction costs dramatically by repeatedly shortening construction schedules.[107] The construction costs for Kwangyang complex were further saved by triggering a 'bidding war' among foreign equipment suppliers who had been suffering from the depression of the world steel industry during the period.[108] POSCO also showed unprecedented speed in absorbing imported technologies.[109] And, as Yoshimori (1991: 296–7) suggests, the strong sense of nationalism, which was evidently expressed in POSCO's motto of 'serving the nation through iron and steel', and shared 'by management of all echelons as well as by the rank and file', would have contributed to the firm-level efficiency, especially during the initial period of POSCO's establishment.

## 7.3 CONCLUDING OBSERVATIONS

The comparison of Japanese and Korean catching-up in the iron and steel industry during the post-war period above broadly fits Gerschenkron's pattern of industrialisation. Latecomers could catch-up with forerunners by establishing 'bigger and bigger' plants with the latest technologies. And in a more backward country, i.e., Korea, the role of the state was more pronounced. This was mainly because functional imperatives from the technological trend of the period were not different from those in the latter half of the nineteenth century.

However, this basic pattern altered with different situational factors in individual countries. In Japan, the structure of oligopolistic competition among *keiretsu* replaced the role played by the German investment banks. Thus, the Japanese iron and steel industry lacked the centre of coordination within the private sector enjoyed by the German investment banks. This coordinating role was undertaken partly by the state in Japan. Moreover, with a rather accidental emergence of 'the developmental state', the role of the Japanese state was more pronounced than in the German case.

In Korea, POSCO's monopolistic position was constrained by the state's overall policy of export promotion and assignment as an instrument for its industrial policy. On the other hand, its international competitiveness was supported by long-term full-capacity utilisation mainly coming from its monopolistic position in the domestic market. Since the Korean iron and steel industry was dominated by this monopolistic firm, an exploration of firm-level factors seems important in explaining the catching-up process, although this is beyond our schema.

The most distinct feature of the catching-up process in the iron and steel industry during the post-war period was the adoption of the global production system. Japan became the first country in history to rise to the position of technological leader in the world iron and steel industry without the benefit of its own natural resources. Exploitation of economies of scale on the global level was an important factor in both Japanese and Korean catching-up. This would be a subject of further research whether or not such globalisation was ubiquitous in other industries.[110] The catching-up process in the semiconductor industry, which will be discussed in the next chapter, shows the process of globalisation in a more accentuated form.

# 8

# USA–JAPAN–KOREA COMPARISON IN THE SEMICONDUCTOR INDUSTRY, 1950–1993

Our historical comparisons in the previous chapters have been focused on one industry, i.e., the iron and steel industry. This is mainly because the iron and steel industry is a typical capital-intensive industry – the kind on which Gerschenkron's schema is based. From the latter half of the nineteenth century, the technological trajectory in this industry was clearly established, moving towards increasing economies of scale. Thus, the functional aspect of catching-up strategy was rather simple: latecomers could gain advantages in international competition by establishing 'bigger and bigger' plants with the latest technologies, although there were some variations across individual countries according to their specific conditions.

Gerschenkron argued that, in the latter half of the nineteenth century, this technological trend was operating on the level of the overall economy, although he acknowledged that his pattern did not fit the cases of the textile and machinery industries, where capital intensity was not a decisive factor in attaining competitiveness (section 3.2). It is beyond the scope of this book to assess whether Gerschenkron's presumption was correct or not in the latter half of the nineteenth century or in a different period. The purpose of this book is more modest. It aims to investigate patterns of catching-up emerging from different technological imperatives, still keeping 'the degree of backwardness' as an organising concept. Chapter 5, in which different patterns of catching-up between Upper Silesia and Rhineland–Westphalia were discussed, was partly devoted to this purpose, still focusing on the iron and steel industry.

This chapter investigates different patterns of catching-up by looking at a different industry, i.e., the semiconductor industry. Technological imperatives in this industry as compared to those in the iron and steel industry are briefly discussed in the first section (8.1). For convenience of comparison we adopt the same comparative setting as in the previous chapter. That is, the catching-up processes of Japan and Korea during the post-war period are compared. In this way, it is hoped, implications for catching-up strategy derived from different technological imperatives shall emerge more clearly.

## 8.1 TECHNOLOGICAL IMPERATIVES IN THE SEMICONDUCTOR INDUSTRY

The semiconductor industry and the iron and steel industry share the same characteristic in that both are basic supply industries of the economy. They provide a wide range of other industries with indispensable components. Their significance is greater than their quantitative share in the economy, mainly due to their great forward linkage effects to the overall economy. However, the former differs from the latter in the following respects.

First, the semiconductor industry is a newly emerging basic industry, whereas the iron and steel industry is currently a mature industry. Of course, the steel industry *was* a newly emerging basic industry in the late nineteenth century.[1] In the industrialisation process of the late twentieth century, however, the importance of the iron and steel industry to national economies has been greatly reduced compared to that of the semiconductor industry.[2] Whilst the growth rate of the iron and steel industry has more or less stabilised, the semiconductor industry is one of the fastest-growing industrial branches in the world economy.[3]

The impact of the semiconductor industry on the overall economy also seems more pervasive than that of the iron and steel industry as 'the micro-electronics (ME) revolution' signifies. For semiconductors not only function as crucial components for other industries, but also embody system functions of other industries.[4] This increasing system-like characteristic of semiconductors is one reason why it is important to view the development of the semiconductor industry in relation with other industries, especially with the other branches of the electronics industry such as computers, telecommunications and consumer electronics. As we will discuss later, the catching-up process in the semiconductor industry has been greatly influenced by progress and necessities in other branches of the electronics industry, and vice versa. In other words, the linkage effect of the semiconductor industry is much greater than that of the iron and steel industry, and this makes it difficult to deal with the development of the semiconductor industry in isolation.

Second, the semiconductor industry is characterised by a more rapid product and process innovation than the iron and steel industry has been. The rapid pace of process innovation is reflected in the quick replacement of capital equipment in the semiconductor industry.[5] And new products have been continually emerging from the transistor, integrated circuits (ICs), microprocessor, electronically erasable programmable read-only memory (EPROM), flash memory, and so on. This rapid product and process innovation makes the competition mechanism in the semiconductor industry quite different from that in the iron and steel industry.

In the latter, final products are not so differentiated and changes in process technologies are not so rapid. The risk involved in initial large-scale investment is therefore not so high. The crucial factor in international competition in the iron and steel industry, therefore, has been initial investment appropriate to exploit economies of scale and state-of-the-art technologies. Once initial investment is made, the major effort of the followers could be concentrated on lowering capital costs, quickening the pace of technological absorption and diffusion, sustaining the learning period, and so on, with lower levels of risk from the possibility of the unexpected emergence of new products and processes. This would be one principal reason why there have been several occasions of technological leadership change from the nineteenth century in the iron and steel industry, whereas such a technological leadership change has not been so evident in the semiconductor industry or the chemical industry.[6]

The different competition mechanism between the two industries is reflected in their difference in capital intensity. The iron and steel industry, especially after the advent of 'the Age of Steel', has been a representative capital-intensive industry. Technological progress in the iron and steel industry has been mostly embodied in physical capital. By contrast, the semiconductor industry lies somewhere between capital-intensive and labour-intensive industrial branches. One pronounced phenomenon in the semiconductor industry is a rapid growth in both labour and capital productivity.[7] It is the intensification of information which brings about rapid growth in capital productivity.

Thus, it is not enough for a latecomer to embark on a large-scale capital investment for catching-up in the semiconductor industry. Catching-up effort through R&D is as important as that through capital investment. Moreover, this catching-up effort, both in R&D and capital investment, is much more a continuous and prolonged one. This technological imperative necessitates embarking on an 'overtaking' strategy at some stage because 'the penalty of being late' is too great in the semiconductor industry, as we shall argue later with the cases of Japan and Korea.

Third, the semiconductor industry, and the electronics industry as a whole, is more characterised by global competition. One reason for this can be attributed to the *spatial separability* of its production process. In the modern iron and steel production, the greatly increased integration of the production process was a major source of technical and economic efficiency. Spatial separation of the production process could be achieved only with a substantial sacrifice of efficiency, at least until the period of Japanese and Korean catching-up.[8] Therefore, the production lines of Japan and Korea remained *within* the national border, although the acquisition of raw materials and marketing of their products were organised on a global scale.

In the semiconductor industry, however, even the production process can be spatially separated without substantially compromising technical efficiency, and, in fact, this can significantly increase economic efficiency. Semiconductor production involves the following five stages:

1 R&D
2 mask making
3 wafer fabrication
4 assembly
5 final testing.

But each stage has different capital and skill requirements. According to J. Henderson (1989: 32):

> R&D, though capital intensive, also demands the application of highly qualified and creative scientists and engineers. Mask making and wafer fabrication are capital intensive also, and require highly skilled engineering and technical labour. However, the latter is distinguished from the former by the fact that it requires, in addition, significant numbers of semi-skilled workers . . . Assembly has been dominantly a labour intensive process given over almost entirely to unskilled labour, although in recent years automation has begun to take place. Final testing, though increasingly capital intensive and requiring significant inputs of technical labour, also requires reasonable amounts of unskilled labour.

Owing to the different capital and skill requirements across different stages of production process, 'a marked polarisation in the skill structure of its labour force' was evident from the initial period of the semiconductor industry.[9] And the movement of the assembly-stage production, which was almost entirely dependent on unskilled labour, by the US merchant semiconductor manufacturers in the 1960s to the offshore regions of lower wages, in conjunction with easy transportability of semiconductors, opened the era of globalised production in the industry. This contrasts with the fact that the followers like Japan and Korea, who lacked natural resources and had a pressing need for export, initiated the utilisation of the globalisation process in the iron and steel industry.[10]

The other reason for the increased globalisation can be found in the rapidity of technological progress in the semiconductor industry. As the existing technologies become obsolescent very rapidly, it is often more profitable for a technology originator to sell its obsolescent technology to followers rather than to remain in production with the technology too long.[11] Thus, technology selling is more prevalent in the semiconductor industry, and the electronics industry in general, than in other industries. The increasing portion of intangible commodities compared to tangible ones in world trade partly reflects the widespread technology trade in the electronics industry.[12]

In this process of global competition, the scope for the autarkic option of catching-up is extremely limited.[13] It seems to us that the dynamics of the semiconductor industry in the East Asian region can be partly attributed to the 'match' between this technological imperative and their export-oriented industrialisation strategies.[14] On the other hand, increased globalisation has opened a new option for entering the industry, that is, 'competitive specialisation' in a narrow range of products or processes. Without heavily investing in the whole production process, a latecomer can start from a less capital-intensive process by exploiting global production networks and global marketing networks, as evident in the initial stage of the electronics in East Asian countries.[15] But the Gerschenkronian option still remains. That is, a latecomer can attempt to catch-up with the forerunner by heavily investing in the capital intensive segment of production. The overall direction of catching-up strategy is not so clearly established in the semiconductor industry as in the iron and steel industry.

## 8.2 JAPANESE ADVANCE IN A NEW INDUSTRY: CONTINUAL STATE INITIATIVES AND INTEGRATED ELECTRONICS FIRMS

### *A convoluted path of catching-up*

The Japanese catching-up with the USA in the iron and steel industry during the post-war period was rather a monotonous 'one-shot' process. There was already a large technological and productive gap between the two countries at the end of the Second World War. But once Japan embarked on catching-up, the gap was continuously narrowed and Japan eventually overtook the USA by a large margin. The USA has never seriously challenged Japan after the relative position between the two countries was reversed. The investment pattern for catching-up also broadly fits the Gerschenkronian one. And the Japanese private sector showed a greater boldness in risk-taking than the state, reflecting the 'moderate backwardness' of the Japanese economy.

By contrast, Japanese catching-up in the semiconductor industry has been a convoluted process, often followed by falling behind. The USA was certainly an indisputable technological leader at the beginning of the industry in the late 1940s. But Japan was initially very quick at imitating and commercialising US inventions.[16] Then, the USA forged ahead with IC technology from the late 1950s. Japan again followed suit, but with several frustrations. Eventually, Japan caught up with the USA in some segments of the semiconductor industry in the late 1970s. However, it could not replace the USA as a dominant technological leader, unlike in the case of the iron and steel industry. Technological leadership in the semiconductor industry is currently divided between the two countries: the USA is stronger in microprocessors and logic, and Japan is stronger in memories and raw materials.[17]

In understanding this convoluted path of catching-up, it is important to draw attention to the linkages between the semiconductor industry and other branches of the electronics industry, especially the consumer electronics and the computer industry. In the 1950s, consumer electronics was a low-tech industry compared to the computer industry and semiconductors, then transistors, were not sophisticated. US leadership in consumer electronics was not so evident and it was overtaken by Japan very quickly. Since the late 1950s, 'virtually all the revolutionary innovations in consumer electronics products . . . have come from Japanese industry', as Gregory (1986: 7) points out. The Japanese initial strength in the production of transistors was closely related to its strength in consumer electronics since it dominated transistor radio and electronic calculator production in the world market.

The US leadership, on the other hand, was initially pronounced in industrial electronics, notably in computers. The main boost for the US computer industry came from the military sector, which took the role of 'creative first user' providing a ready premium market for new products.[18] Relying on this military demand, IBM pioneered the transistorised computer in 1955 and retained its dominant position in the world well into the 1970s. The US semiconductor industry forged ahead from the late 1950s with the growth of its computer industry. In responding to the military demand for high-performance products and miniaturisation, the US producers pioneered silicon-based ICs. By contrast, Japan and European countries were falling behind by focusing their R&D efforts on the improvement of germanium-based transistors used mainly in the consumer electronics products.[19]

Japan attempted to catch-up with the USA in the computer industry almost from the beginning of the post-war period.[20] It intensified its catching-up effort in the semiconductor industry in the 1960s, since it recognised that semiconductor technology was essential in narrowing the gap in the computer industry. Japan imported the US semiconductor technologies heavily all through the 1960s. But the USA continually forged ahead with introduction of new products in the computer and semiconductor industry.[21] IBM became the largest producer of advanced ICs in 1965 purely on the strength of its own internal demand.[22] Some Japanese semiconductor manufacturers even 'accused US firms of "dumping" in Japan' in the early 1970s.[23]

These continual frustrations in catching-up with the USA in the 1960s provided a main impetus for accelerating Japanese catching-up effort in the 1970s. Moreover, the Japanese strength in the consumer electronics was beginning to be challenged by the USA in the early 1970s. Japanese consumer electronics producers had been relying mainly on Japanese-made transistors. But American companies began competing, offering products incorporating advanced ICs. Owing to such competitive pressure, the Japanese share of the domestic US calculator market fell sharply to 25 per cent in 1974 from around 85 per cent in 1971. Leading Japanese calculator producers had to 'enter into long-term contracts with [the US semiconductor producers like] Rockwell, GI, Texas Instruments, and Fairchild for calculator chips and technical assistance to close the technology gap'.[24]

Japan appeared to narrow the technology gap with the USA in the computer industry with its accelerated catching-up effort. By 1974, 'Fujitsu introduced a clone of the extremely successful IBM 370 series, Fujitsu's 370 M, which was faster than IBM's machine'.[25] Advancement was also achieved in large-scale integration (LSI) production. According to Borrus, Millstein and Zysman (1983: 208), 'Japanese semiconductor-computer firms had developed a significant LSI capability, and by 1976 they dominated their domestic market in all but the most advanced IC devices'.

But, a new challenge again came from the forerunner. One significant market development was 'the introduction by US computer companies of low-cost, LSI-based plug-compatible mainframes (PCM)' in 1975.[26] It was only a year before that Fujitsu 370 M, which was based on IC technology, came to market. Moreover, IBM's plan to develop a new generation of computers based on very large-scale integration (VLSI) technology posed a further threat to the Japanese catching-up strategy.[27] Japanese intention to 'enter international competition in LSI-based mainframes had been rendered moot' by these two developments in the international market.[28]

At this time, Japan adopted a different strategy, that is, an *overtaking strategy*. All the previous efforts were based on catching-up strategy: they were aimed at narrowing the existing technology gap with the forerunner. But the new strategy, which was manifested in the well-known VLSI Project (1976–1980), was geared 'at developing semiconductor technology for the *next* generation of computers'.[29]

This had mainly to do with the different technological imperatives in the semiconductor industry as compared to those in the iron and steel industry. In the latter, it was not an urgent task for latecomers such as Germany or Japan to develop by themselves state-of-the-art technologies like Thomas basic process or BOF; it was sufficient for them to exploit new technologies originated from their forerunners.

But the rapid pace in product and process innovations set by the US semiconductor producers had caused continual frustrations to the Japanese catching-up effort as discussed above. Moreover, the risk involved in investment with existing technology was greatly increased since capital requirements for developing new generations of computers and semiconductors was increased dramatically and the pace of capital replacement became more rapid from the 1970s. Thus, 'if Japanese computer companies wanted to break into the international market for computers in a big way, they were going to *have to* do so on the basis of the next generation of semiconductor-computer technology – VLSI'.[30]

The VLSI Project included a broad range of R&D related to the VLSI technologies, reflecting the fact that state-of-the-art technologies in semiconductors are enormously complicated. But the most significant advance was made in dynamic random access memory (DRAM) production technologies. Japanese producers commercialised 64K DRAM almost at the same time as their US competitors did.[31] By the end of 1981, they captured 70 per cent of the international market for 64K DRAM.[32] Japan became the world leader in a segment of the semiconductor industry for the first time since IC had been introduced in the late 1950s. It would be surely an exaggeration to attribute this advancement of the Japanese semiconductor industry solely to the VLSI Project. But it cannot be denied that, during the period of the VLSI Project, the Japanese semiconductor industry was given critical momentum for competing with its US counterpart on a more equal basis.[33] From the mid-1980s, Japan could dominate the world RAM market. It rose to become the largest supplier of application-specific ICs (ASICs) in 1987 and significantly narrowed the gap with the USA in microprocessor production by the late 1980s.[34] Japan began dominating the semiconductor equipment and material markets from the mid-1980s.[35]

### *Functional pattern of the Japanese catching-up*

We emphasised above that the rapid pace of product and process innovation in the semiconductor industry increased the risk involved in the initial large-scale capital investment on the part of latecomers. What should be noted, however, is the fact that the Japanese catching-up was most pronounced in the capital-intensive segment of the semiconductor industry, in particular in DRAM production. This has to do with the establishment of the technological trajectory in the industry. Two segments in the industry, i.e., RAM and microprocessors, began to show a clearly technological trend towards mass production and a dramatic increase in capital intensity in the 1970s.[36] Japan opted for concentrating their initial catching-up effort on DRAM because its design was 'regular and conceptually simple' as compared to that of the microprocessor.[37]

In this respect, the Japanese catching-up strategy in the semiconductor industry is a microcosm of Gerschenkron's pattern. According to Gerschenkron, a producers' goods industry, that is, a capital-intensive industry among various industries within the economy, was an easier entry for catching-up in the latter half of the nineteenth century. Likewise, the Japanese experience shows that DRAM, a capital-intensive segment among various segments within the semiconductor industry, was a promising point of entry for the latecomer to make inroads in the state-of-the-art semiconductor technology.

Unlike Gerschenkron's original schema, however, the catching-up strategy in the capital-intensive segment of the semiconductor industry was not simply a matter of establishing 'bigger and bigger' plants with latest technologies. First of all, the mobilisation of financial resources had to be a continuous rather than a 'one-shot' process because of the rapid capital replacement in the industry. Second, since capital replacements were inevitably accompanied by new products and new processes, catching-up effort through R&D became more important. R&D was essential to counter uncertainties from the continual emergence of new products and processes, and it was all the more essential in adopting an 'overtaking strategy'.

Japan fared better than the USA in financing the continuous investment, mainly through its institutional structure of the *keiretsu* system and the developmental state, as shall be discussed later. Japanese investment increased more rapidly than that of the USA all through the 1970s.[38] Continued capital investment by Japan was most remarkable during the period of recession after the first oil shock. The US semiconductor manufacturers restrained their investments in the development of new technologies and production capacity. By contrast, Japanese producers continued a high rate of investment. Whereas the average ratio of capital spending to sale of the US producers remained only 12 per cent during the period 1973–1978, that ratio for the Japanese producers was 17.8 per cent during the same period. In 1983, the 'total capital spending' of Japanese semiconductor makers surpassed that of the US makers.[39]

The Japanese R&D effort also shows a distinctive pattern. In order to overtake its forerunner, its R&D investment was focused narrowly on process technologies like electron-beam lithography. It was simply impossible to overtake forerunners in all the technologies from the beginning. Although the VLSI Project produced more than 1,000 patents, they were not, in essence, technological breakthroughs. Instead, Japanese products excelled in producing better quality products through incremental innovations and the adoption of 'a more conservative design and a better ceramic package'. Japanese firms also significantly lowered the defect ratio by increasing cleanliness in the production process.[40]

In the catching-up process in the iron and steel industry after the advent of the Age of Steel, this kind of unbalanced focus within the industry did not appear because its production process was tightly integrated and products were differentiated only at the later stages of production process. In the semiconductor industry, however, products are segmented, and capital and skill requirements are rather different across those segments. Here, Japan adopted an *unbalanced growth strategy within the semiconductor industry*. It started its catching-up from a narrow segment of semiconductor products and technologies appropriate to utilising its relative strength in capital mobilisation and mass production, and then extended its catching-up gradually into more sophisticated products and technologies. Basically the same strategy was adopted by Korea, an 'extremely backward' country, as we shall see later (section 8.3).

### *Continual state initiatives*

In the case of the iron and steel industry, we emphasised the initiatives taken by the Japanese private sector. But it is difficult to draw a clear line between the private and state initiatives in the Japanese catching-up in the semiconductor industry. In the 1950s, when the semiconductor industry was based on transistors, the Japanese private sector initiated catching-up without any particular support from the state. But the Japanese state intervened heavily in the computer industry from the beginning. It began to intervene heavily in the semiconductor industry after the introduction of ICs and sometimes initiated the catching-up process, as can be seen in the case of the VLSI Project.

This was partly because the semiconductor industry was quite a young industry with lots of uncertainties. In fact, even in the US, the forerunner country, the initial market and direction of technological progress in the computer and semiconductor industries were led by the state. The role of the state as a 'creative first user' seems crucial during the 'formative stages of growth' of an industry regardless of the degree of backwardness of a country, because uncertainties in the industry are simply too great.[41] This fundamental uncertainty in a new industry would explain the reason why the initial research of the computer industry in Japan was undertaken by the public sector in the 1950s.[42]

However, the Japanese state's heavy involvement in the computer and semiconductor industries, even after the formative stage of the industry, should be understood in view of the combination between the Japanese state's ambitious catching-up strategy and constant frustrations in catching-up, at least up until the mid-1970s. If the Japanese computer and semiconductor industries had been able to catch-up with their US counterparts as successfully as the consumer electronics industry or other industries had done, there would not have been so many government initiatives for narrowing the technology gap.[43] On the other hand, if the Japanese state had not been so ambitious as to catch-up with the USA in broad segments of the electronics industry, it might have left the lagging Japanese computer and semiconductor industries until its private sector should develop enough to initiate catch-up by itself.

It seems important here to pay attention to the element of Japan's 'moderate backwardness'. As will be discussed later (section 8.3), this ambitious catching-up strategy from the beginning was simply not available to Korea because of its 'extreme backwardness'. Japan had, however, already accumulated technological and organisational capability, especially through the development of the telecommunications and electrical power industries before and during the Second World War, to embark on a broad-front catching-up from the beginning of the semiconductor and computer industries.[44]

Backed by its moderately developed economy, the Japanese state could intervene in the industry more pervasively. A tight protective system was maintained until the Japanese semiconductor industry gained international competitiveness.[45] A quota system was devised especially to block the imports of complex US ICs assembled in East Asia.[46] Moreover, the Japanese state tightly controlled technology inflows and hardly allowed direct investment by the USA MNCs.[47] 'Buying (using) Japanese' policy also contributed to preserving a large portion of domestic market for the Japanese producers.[48]

As a result of these various protective measures, for instance, the US firms could penetrate only 20 per cent of the Japanese market for ICs in 1975 while their share was 98 per cent in the USA and 78 per cent in Western Europe.[49] The developmental sequence of the Japanese semiconductor industry was to acquire competitiveness in the protected domestic market and then to move on to the export market. In this respect, the overall policy direction of the Japanese state in the semiconductor industry can be termed as import-substitution (IS) – *then* – export promotion (EP). This pattern of development was not available to Korea because of its 'extreme backwardness'. Korea had to adopt an EP-*cum*-IS policy as will be elaborated later (section 8.3).

One distinctive role of the Japanese protective system in the semiconductor industry was that it was employed as a means of acquiring foreign technologies. As technological progress was rapid in the industry, it was often profitable for the technology originator to license its technology if marketing opportunities for its products were blocked. Japanese tight protection in its domestic market therefore led US semiconductor firms 'to license their technology to Japanese firms as a means of generating residual earnings in a market to which access has been difficult at best'.[50] This contrasts with the case in the iron and steel industry, where the protective system was not related to technology transfer.

If we compare the Japanese state's intervention with that of its forerunner, i.e., the USA, it had a stronger 'organisational direction' as a latecomer.[51] As in the case of the iron and steel industry, the Japanese developmental state provided its semiconductor industry with various macro-level stimuli, among which the 'easy money' policy of the state was a very important measure.[52] This easy money policy enabled a more stable supply of capital to its industry than the US system, as was the case in the iron and steel industry. The US firms' heavy reliance on internal financing was unavoidably susceptible to business fluctuations.[53] By contrast, the Japanese firms could maintain a stable rate of investment supported by long-term commitment from the banking sector.[54] This stable availability of capital in Japan was reinforced by the *keiretsu* system and the structure of integrated electronics manufacturers, as we will discuss below.

Also, as in the iron and steel industry, the Japanese state's industrial policy contributed to catching-up in the semiconductor industry. The semiconductor industry, and the electronics industry in general, received the state's support as one of the strategic industries, especially from the 1970s. There were, however, some differences in the application of the state's industrial policy as compared to that in the iron and steel industry.

The first was related to the stronger need for R&D in the semiconductor industry. One conspicuous feature of the Japanese state intervention in the semiconductor industry was coordinated research. As commercial interests in the computer industry grew from the late 1950s, the Japanese state embarked on a series of ambitious research projects for catching-up in the computer-related industries: the FONTAC Project (1962–1966), VHSCS (Very High Speed Computer System) Project (1966–1972), the Mainframe Computer Project (1972–1976), the PIPS (Pattern Information Processing System) Project (1971–1980), the VLSI Project (1976–1980), and so on.

Despite the fact that individual projects differed from one another in their specific emphases, the objective of the coordinated research projects was generally twofold: to economise on R&D resources by reducing duplicative research and to hedge risks at the economy level through the division of specialisation in research. The latter ensured that the 'Japanese system as a whole would avoid being seriously weakened as a result of having backed the wrong horse'.[55] The former was most pronounced in the VLSI Project which was undertaken by establishing joint laboratories. Although individual firms initially opposed MITI's idea of joint labs and were eventually 'forced' by MITI to join the project, the project produced a great saving effect at the level of the industry and individual firms.[56] With savings from the joint research, individual firms could concentrate more on capacity expansion.[57]

Another role of the state which is distinguished from that in the iron and steel industry was technology forecasting. As the semiconductor industry was a young industry and the pace of technological progress was rapid, the state's overall guidance of the industry could play an important role in reducing uncertainties and thereby promoting investment. The beginning of the Japanese computer industry was therefore initiated by public institutions. Moreover, the Japanese government identified electronics as a growth sector early in 1957 and enacted 'the Provisional Measures Law Concerning the Promotion of the Electronics Industry' (1957–1969).[58] The 'turning point' for transistorised TV production was made by the government's push for 'going all solid state TV'.[59] And MITI proposed a 'knowledge-intensive growth vision' in 1971 prior to the first oil shock, and this was understood by many observers to have contributed to mitigating the shock.[60]

### *The integrated electronics firms*

We have emphasised the importance of the *keiretsu* system in the Japanese catching-up in the iron and steel industry (section 7.1). Here, the major function of the *keiretsu* lay in the provision of capital. This synergy mainly came from the structure of horizontal *keiretsu*, which is centred around big city banks. Major Japanese semiconductor producers are certainly members of horizontal *keiretsu*. But their degrees of involvement in the horizontal *keiretsu* groupings vary greatly across individual firms. Whilst some of them have a close link with a horizontal *keiretsu* grouping (NEC with Sumitomo Group, and Mitsubishi Electronics with Mitsubishi Group), many of them have only loose ties with horizontal *keiretsu* groupings (Fujitsu with DKB Group, Toshiba with Mitsui Group, Matsushida with Sumitomo Group), and one of them, Hitachi, has multiple loose connections (with Fuyo Group and Sanwa Group).[61] Japanese semiconductor producers in general surely benefited from this horizontal *keiretsu* grouping. Some of them, especially those who had a close link with the *keiretsu* grouping, received critical support from their *keiretsu* banks in their continued investment.[62]

A more distinctive organisational feature of the Japanese semiconductor industry, however, is that all of the major semiconductor producers are actually integrated electronics manufacturers whose businesses cover not only semiconductors but also consumer electronics, computer, telecommunication, and so on.[63] And many of them, Hitachi, Toshiba, Sony and Matsushida, grew into independent *vertical keiretsu* which were formed mainly through backward and forward linkages of the interrelated industries without mediation of financial institutions. The share of semiconductor sales to the total sales of Japanese electronics firms is therefore invariably very low.[64]

This closer integration of semiconductor production with other branches of electronics would be mainly attributed to the system-like characteristic of the semiconductor. In the case of the iron and steel industry, Japanese producers were rarely integrated into their upstream or downstream industries. Horizontal *keiretsu* grouping seems to have been enough to increase the synergy between the iron and steel industry and other related industries and financial institutions.[65] But the system-like characteristic of the semiconductor industry considerably increased the importance of close interaction between it and its downstream industry. In Rosenberg's (1982: 135) terminology, the 'high degree of systematic complexity' of the semiconductor industry generally increased the importance of 'learning-by-using'.

In the USA, however, the integration of electronics producers was blocked by the anti-trust law which prohibited system producers like IBM and AT&T from entering merchant semiconductor manufacturing.[66] Without legal restrictions, but with the existing *keiretsu* grouping, the Japanese producers could exploit the synergy between the sectors better than their US counterparts. In the initial stage of development, consumer electronics and semiconductor production for consumer electronics were a 'niche market' for Japanese firms.[67] As far as international competition was concerned, it was more promising for Japanese firms to start from concentrating on those segments. Then they integrated or diversified into more technologically sophisticated areas. Japanese semiconductor producers' high degree of consumer electronics orientation can be understood in this context.[68]

One critical synergy from this broad integration lay in close and systemic user–producer relations. As discussed above, a main impetus for upgrading the level of Japanese semiconductor technology came from its computer sector. The need for supplying sophisticated ICs for their computer divisions facilitated the development of the semiconductor divisions within the integrated electronics firms. On the other hand, the semiconductor divisions in turn facilitated the development of other divisions.[69]

The other important synergy came from demand-side promotion within the integrated electronics firms. For instance, about 21 per cent of the value of semiconductor production was consumed internally in the ten largest producers in 1980.[70] The existence of this ready internal market provided Japanese firms with a minimum stable demand for their new products and thereby enabled faster returns on their investments.[71]

Here again, we can detect an element of 'moderate backwardness' of the Japanese economy. As we will discuss later (section 8.3), the Korean semiconductor industry developed relying much less on internal demand and technological linkages, even with the *chaebol* grouping, mainly because its electronics industry was 'extremely backward'. By contrast, Japanese producers could rely on themselves for the initial market and for technological support because their electronics industry was developed to some extent. This was one reason why Japan could pursue IS-*then*-EP strategy.

The structure of integrated electronics firms also helped the development of the Japanese semiconductor industry financially. Within the integrated electronics firms, semiconductor divisions 'not only have access to low-cost capital but also can take indirect advantage of tax incentives and other special benefits extended to other divisions of the company'.[72] Moreover, incomes arising from the non-semiconductor section could be easily diverted to semiconductor investment. The advantages of such internal subsidisation and finance were most pronounced in the periods of recessions. Undergoing recessions both in the mid-1970s and the mid-1980s, the pace of Japanese catching-up was more accelerated, as we saw above. If Japanese semiconductor manufacturers had been merchants like their US counterparts, their sustaining of high investment rates would have been difficult.[73]

## 8.3 KOREAN CATCHING-UP IN A HIGH-TECH INDUSTRY: CHANGING CONFIGURATION BETWEEN THE STATE AND THE CHAEBOLS

In the Japanese catching-up in the semiconductor industry, public and private initiatives were intermingled. This was because Japan pursued a catching-up strategy in broad segments of the electronics industry from the beginning and because the Japanese private sector was successful in catching-up in some segments like consumer electronics, but not quite so in other segments like semiconductors and computers. By contrast, the Korean catching-up in the semiconductor industry was a rather linear process, starting from simple assembly based on low wages and gradually moving towards more technologically sophisticated processes as the Korean economy developed. It seems that the development of the Korean semiconductor industry can be divided into three phases as follows:

1 initial beginning as a pure export industry through foreign direct investment (FDI) and state initiative (1966–1975);
2 transitional period (1975–1980);
3 take-off and import-substitution through private sector initiative (1981–).

*Phase 1: 'enclave' development as an export platform of MNCs*

In the iron and steel industry, Korea faithfully followed the Japanese catching-up strategy from the beginning. In the semiconductor industry, however, it was simply impossible for Korea to adopt the Japanese strategy from the beginning. It could not attempt catching-up in high-tech sectors like the computer and semiconductor industries at least until the late 1970s. Its initial catching-up effort in the electronics industry was concentrated only on consumer electronics. This was a reflection of Korea's 'extreme backwardness' compared to Japan's 'moderate backwardness', on the one hand, and of different technological imperatives in the electronics industry compared to those in the iron and steel industry, on the other.

These differences were clearly reflected in the 'dual' structure of the Korean electronics industry.[74] In the 1960s, the Korean state pursued a very strict import-substituting policy in the electronics industry as in other industries. It banned the imports of finished consumer electronics products like radios, monochrome TV sets, and so on.[75] Its domestic market for the consumer electronics products was exclusively secured for local producers.

In the beginning, the Korean consumer electronics was simply 'an assembly operation of foreign components and parts with equipment purchased from overseas'.[76] And later, the import-substitution of components was attempted by allowing the imports of parts and components only if 'the user provided evidence of inferior or insufficient local supply'.[77] One could say that, in the Korean consumer electronics sector, Hirschman's (1958; 1968) gradual import-substitution through backward linkage effect was pursued in the protected domestic market.

It should be noted, however, that this import-substituting policy was pursued simultaneously with export promotion. The electronics industry was one of the industries which showed a trend towards spatial separation of production processes during the post-war period. It was therefore convenient for Korea, which had, like other developing countries, a comparative advantage in low-wage labour, to promote the assembly of electronics products as an export item for earning hard currency.

Some segments of the electronics industry, like radios and TV sets, were developed for export supported by the protected domestic market. But other segments were developed solely for export without significant linkages with domestic demand. For instance, electronic calculators, tape recorders and electronic digital watches were initially assembled almost exclusively for the export market.[78] Reflecting this export-oriented development, around 51 per cent of consumer electronics products were exported by 1972. The insulation from domestic demand was more pronounced with electronic parts and components like semiconductors, capacitors, resistors, and so on, for which the ratio was around 82 per cent in 1972.[79] The initial stage of the Korean electronics industry was therefore characterised by the dual structure consisting of a segment which had a linkage with domestic demand and a segment of 'enclave' sustained in the network of international specialisation without a significant linkage with domestic demand.

The initial development of the Korean semiconductor industry should be viewed in the context of this dual structure of the electronics industry. The beginning of the Korean semiconductor industry in the middle of the 1960s was a typical 'enclave' development principally aimed at earning foreign currency. This enclave characteristic was much stronger in the semiconductor industry than in other sectors of the electronics industry as it was initially developed by 100 per cent foreign direct investment.[80] All the equipment, raw materials and intermediate goods were imported and almost all the semiconductors processed in Korea were exported, as Table 8.1 shows.[81]

As pointed out before (section 8.2), Japanese firms could acquire and utilise advanced technologies through licensing or indigenous technological effort mainly because of their 'moderate backwardness'. This option was, however, not available to Korea due to the combination of MNCs' strategy and Korea's 'extreme backwardness'. MNCs' initial strategy was simply to exploit low wages in developing countries without transferring their advanced technologies. And the Korean state had neither the intention nor the ability to negotiate with MNCs for technologies, or to force its private or public sector towards technological catching-up. It induced FDI mainly for earning foreign currency. The Korean semiconductor industry started entirely as an 'export platform' for US and later Japanese semiconductor manufacturers.

It is important to note, however, that the case of the semiconductor industry was exceptional in the overall process of Korean industrialisation. The Korean state has heavily intervened in the introduction of foreign

*Table 8.1* The share of export to production in the Korean semiconductor industry

| *Year* | *1966* | *1971* | *1976* | *1981* | *1986* | *1987* | *1988* | *1989* |
|---|---|---|---|---|---|---|---|---|
| % | 100 | 95.9 | 94.6 | 96.0 | 94.1 | 88.7 | 86.4 | 89.4 |

*Source*: Yoon (1990: 70, table 1); Kim, Hong and Lee (1992: 245, table 4.6)

capital and technologies.[82] In fact, the beginning of the semiconductor industry in Korea in 1966, which had to deal with Fairchild's demand for 100 per cent ownership and the right to sell its products in the domestic market, was 'the first legal collision' concerning the Law for Importation of Foreign Capital which did not allow 100 per cent foreign ownership.[83] For the Korean state, allowing an exceptional case was better than strictly applying its regulation for its industrial development because its economy then had neither the technologies nor market for the high-tech industry. In this respect, the beginning of the semiconductor industry in Korea shows a limitation in the role of the state constrained by external conditions and internal capability, although it was a highly interventionist state on the level of the overall economy.

Nevertheless, it is also important to note that it was the state which initiated the beginning of the semiconductor industry in Korea. It was officials of the Ministry of Commerce and Industry (MCI) who first contacted Fairchild's managers in 1965 and persuaded them to invest in Korea.[84] In 1970 the state also established 'free trade zones' to induce foreign investment in the semiconductor and other electrical components, especially from Japan.

The role of the Korean state was larger than was immediately apparent within the semiconductor industry. The state contributed significantly to the future development of the semiconductor industry by promoting its electronics industry.[85] The electronics industry was given a strategic importance during the HCI programme in the 1970s. The meagre emergence of some indigenous semiconductor manufacturers in the early 1970s can be understood within the general environment of the state's promotion of the electronics industry. Anam, which later became the largest semiconductor assembler in the world, was set up in 1970 as a subcontractor for foreign semiconductor producers.[86] And Goldstar and Samsung also began semiconductor assembly in the early 1970s, though on a very small scale, mainly for their consumer electronics business.[87]

### *Phase 2: transitional period*

In Korea, the imperatives for technological catching-up in the semiconductor industry arrived only with the strengthening of its indigenous electronics manufacturers, mainly those of the consumer electronics sector. The production and export of electrical parts and components were still dominated by foreign firms even up until the mid-1970s.[88] But local consumer electronics manufacturers like Goldstar, Daehan, and Samsung were firmly established in the 1970s.[89]

They upgraded their technological levels gradually, increasing local contents of parts and components and then moving on to more advanced products. Various technology licence or joint venture agreements were concluded for this purpose. They could also penetrate the international market with successive waves of consumer electronics products like radios, electrical calculators, TV sets, and so on. On the other hand, foreign electronics firms, who had been relying mostly on low-wage unskilled labour, began retreating from Korea or reducing their shares in joint ventures in the 1970s as the wage level of Korea was rapidly increased during the HCI programme.[90]

Broadly speaking, the successful development of the consumer electronics industry brought about the need for 'capital deepening' as in other industries.[91] But the degree of the urgency for capital deepening seemed stronger in the electronics industry than in other industries. Most semiconductors and other components for Korean consumer electronics products had been imported from Japan. While Korean producers concentrated on low-tech products, relying mainly on low wages, this dependence on crucial components of the Japanese producers was not such a serious problem. But Japanese semiconductor producers were also integrated electronics firms highly oriented to consumer electronics. Once the wage level was increased and it therefore became imperative for the Korean producers to begin competing directly with the Japanese in more sophisticated consumer electronics products, the dependence on Japan for semiconductor and other components was perceived as no more than dependence on a potential competitor.

This urgency was seriously felt both by the Korean state and by the private sector from the mid-1970s. For the state, the development of the semiconductor industry was recognised as a crucial factor in the success of its ambitious HCI programme, since the electronics industry was one of the designated industries of the programme. The state announced a series of promotion measures for the semiconductor industry, which we will discuss later in detail. For private electronics firms, the entry into the semiconductor business became imperative from the viewpoint of long-term corporate strategy and they began moving to semiconductor wafer fabrication from simple assembly in the mid-1970s.

It is difficult to assess the precise balance between the role of the state and that of the private sector during the latter half of the 1970s. But there seemed to exist a broad division of labour between the state and the private sector. The private sector, which was led by *chaebols*, focused its effort on the fields directly related to its consumer electronics business. For instance, Samsung, the head-starter, was operating its semiconductor business on a small scale essentially for internal supply for its consumer electronics products.[92] The actual investment and R&D during the period was still very small in scale. But Samsung and Goldstar showed a strong commitment to the semiconductor industry by 'increasing their investment in the semiconductor sector which had continuously recorded deficits even when they were reducing investment in other sectors during the period of severe recession [from the late 1970s]'.[93]

On the other hand, the state's effort was concentrated more on the overall R&D activities as well as providing a broad institutional framework for the development of the local semiconductor industry.[94] In 1975, semiconductor technology was designated as the most important research item of the Korea Advanced Institute of Science and Technology (KAIST), a principal public R&D institute. The Korea Institute of Electronics Technology (KIET), a new research institute specialising in semiconductors and computers, was established in 1976 as a public institution. Its aims were to gain 'the first-stage experience in R&D on new technologies and to produce experienced researchers and production workers'.[95]

The pioneering role of KIET seems to deserve special attention. It functioned as a collaborative research institute between the state and private sector aimed at providing instant manufacturing technologies for private firms.[96] Thus, KIET had not only basic research facilities, but also 'pilot production facilities' for memory chips and computers.[97] With these facilities, KIET helped private firms overcome problems arising from production processes and succeeded in developing 32K and 64K ROMs in the early 1980s, when none of the private firms yet made ROM or RAM.[98] KIET also established a liaison office in Silicon Valley in 1978 and assisted private firms in gathering technological information, and recruiting Korean scientists employed in the USA.[99]

Another aspect of state intervention in the semiconductor industry distinguished from that in the iron and steel industry was the method of intervention. In the case of the latter, the Korean state significantly increased its involvement in the industry when its catching-up effort began in earnest. But it is an important fact that whenever the Korean government attempted to promote the electronics industry it relaxed the existing regulations on foreign technology and capital.

As mentioned above, the state allowed 100 per cent foreign direct investment in the semiconductor industry as an exceptional case in 1966. The establishment of free trade zones itself was to provide an exceptional space for foreign firms to conduct their business in Korea 'free' from the existing government regulations. And when the state strengthened its effort to promote import-substitution in the electronics industry in 1975, one significant step was 'to relax regulations concerning foreign direct investment in the electronics industry for the purpose of advancing technological level'.[100] A similar step was again taken in 1983 as the state 'revised its regulation of the introduction of foreign capital and designated the advanced electronics sector as *completely open to FDI*'.[101]

These relaxations of regulations on capital and technology inflows were not in the least a relaxation of the Korean state's import-substituting effort. In fact, import-substitution remained a strategic pillar of the state's intervention in the economy up until the late 1980s.[102] However, a different method of import-substitution had to be adopted because of its relative backwardness. As we pointed out before (section 8.2), the Japanese state could pursue import-substituting policies by tightly controlling technology and capital inflows and by maintaining a protective domestic market in the semiconductor industry, mainly because it was backed by its moderately backward economy. For Korea, which lacked enough of a technological base and domestic demand, allowing freer inflow of foreign technologies and capital was a better way of facilitating technological progress in the high-tech industry.

Our previous discussion on technology transfer (section 2.1) is relevant here. Generally speaking, the actual pace of technological progress in the latecomer is decided by the combination of the volume of technology inflow and the local technological effort at assimilating imported technologies. If the inflow of advanced technologies is minimal, even the most strenuous effort at assimilation does not bring about a satisfactory pace of technological progress in the latecomer. The Korean state's continual relaxation of its regulations on foreign capital and technology would be understood as an effort at facilitating technological progress of local industry by increasing the volume of technology inflow when the possibility of technological upgrade relying on local capability and domestic market was limited.

However, the concerted effort between the state and the *chaebol* in the 1970s did not bring about an immediate result. Most attempts by new entrants to the industry were frustrated. In the existing manufacturers, the assembly production was still dominant despite some technological advances mentioned above. This limited achievement during the period can be still attributed mainly to the difficulty in securing and assimilating advanced technologies.[103] Korea simply required more time for its economy to develop and a new competitive structure to emerge.

### *Phase 3: full-fledged catching-up by the initiative of the chaebol*

The take-off of the Korean semiconductor industry occurred in the middle of the 1980s with aggressive investment by major *chaebols*. The Korean producers started their catching-up by concentrating their effort on DRAM production just as their Japanese counterparts had done about a decade before. Major Korean producers began producing 64K DRAM from 1984 and 1985.[104] At first, Korea was behind the world leaders by 6 years in development and by 4 years in production of DRAM. But the time lag was continuously reduced as new generations of DRAM were developed. In 4M DRAM, Korea narrowed the time lag in production to only one year and it began producing 16M DRAM without any significant time lag with the forerunners, as Table 8.2. shows.

In 1991, Samsung became the world's largest producer of 1M DRAM. It became the world's largest producer of all DRAM products by 1992 and the world leader in memories by 1993.[105] Hyundai and Goldstar were ranked as the tenth- and eleventh-largest producer of DRAM respectively in 1993. Korea's share of world DRAM production reached 20 per cent in 1990 and 23.6 per cent in 1993. And its total share of memories was increased to 13 per cent in 1990 and 17.9 per cent in 1993.[106] Korea's catching-up in other segments of the semiconductor industry is not so impressive as that in memories. As far as DRAM production is concerned, however, it is safe to say that Korea has already reached the forefront.[107]

The Korean state played a significant role in this rapid catching-up in the semiconductor industry during the period. It designated the semiconductor industry as a leading sector in the 1980s.[108] And various promotional measures were followed.[109] The Korean state also organised collaborative research projects for the development of new generations of semiconductor from the development of 4M DRAM in 1986.[110] It was reported that Samsung's acquisition of design and manufacturing technology of 4M DRAM in 1988 was significantly helped by collaborative research.[111]

The initiative for catching-up during the period, however, mainly came from the private sector. Yoon (1990: 107) argues, for instance, that 'the state's actions had to do more with *ex post* recognition of the situation progressed by the private sector than with its initiative of signalling inducements for its private sector to get moving'. One impetus for this private sector initiative was already at work. As we pointed out above, the

*Table 8.2* Time lag of DRAM development and mass commercial production

| | | *64K* | *256K* | *1M* | *4M* | *16M* |
|---|---|---|---|---|---|---|
| | World | 1977 | 1980 | 1983 | 1986 | 1987 |
| *Development* | Korea | 1983 | 1984 | 1986 | 1988 | 1990 |
| | lag (year) | 6 | 4 | 3 | 2 | 3 |
| | World | 1980 | 1982 | 1985 | 1989 | 1992 |
| *Production* | Korea | 1984 | 1985 | 1987 | 1990 | 1992 |
| | lag (year) | 4 | 3 | 2 | 1 | 0 |

*Source*: Chu (1992: 43, table 3.11)

existing electronics firms needed semiconductor technologies from the latter half of the 1970s as their competitiveness based on low wages was rapidly being eroded.

A more important impetus which emerged in the 1980s was the growth of *chaebols*. Although *chaebols* had existed from the early 1950s, they experienced a substantial expansion during the HCI programme in the 1970s.[112] Once securely established, they searched new businesses for further growth in the 1980s. The semiconductor industry was regarded as one of those businesses. As the ME revolution signifies, it would be very difficult to enter other promising businesses like telecommunications, computers and aircrafts, or to further develop existing businesses like automobiles and machinery, without semiconductor technologies.[113] Moreover, the long-term prospect of demand growth in the semiconductor industry itself was promising, though short-term entry costs might be enormous.

Thus, existing wafer manufacturers like Samsung and Goldstar increased their investment significantly both in facilities and R&D from 1981, even with accumulating losses. The entry of Hyundai in particular was related to its strategic concern as a business group. It entered the semiconductor industry without prior experience in the electronics industry, but embarked on ambitious investments which were even bigger than those of Goldstar, then the second largest electronics manufacturer in Korea.[114] The Daewoo Group, then the fourth-largest *chaebol* which had started consumer electronics business only in the late 1970s, also entered the semiconductor industry by acquiring semiconductor and consumer electronics divisions of Daehan. Hence the emergence of oligopolistic competition: four major *chaebols* began competing fiercely with each other in one industry.

The private sector initiative in the 1980s can be clearly detected from the structure of financing for semiconductor investment. As one company's (probably Samsung) case illustrates (Table 8.3), internal financing and foreign financing took the major portion of total finance for investment whereas external financing through domestic banks was significantly reduced. Major *chaebols* began heavily utilising the domestic capital market to raise their investment funds from the mid-1980s.[115] They also turned to the international capital market for investment funds for their semiconductor business.[116]

What should be noted is that foreign financing in this period was qualitatively different from that in the 1970s. Whereas foreign loans were raised by the state initiative and its guarantee in the 1970s, *chaebols* now raised foreign loans with their own credit. The Korean state in the early 1980s was actually restraining bank loans to *chaebols* responding to the growing domestic criticism of the *chaebol*.[117] In this situation, the international capital market functioned as an alternative source for investment funds for the *chaebol*.[118]

*Table 8.3* Financing of investment in a semiconductor firm (Won 100 million, %)

| | *Domestic source* | | | |
|---|---|---|---|---|
| *Year* | *Internal* | *External* | *Foreign source* | *Total* |
| 1983 | 83 (31) | 113 (42) | 73 (27) | 268 |
| 1984 | 270 (27) | 204 (20) | 526 (53) | 1,000 |
| 1985 | 203 (19) | 145 (14) | 695 (67) | 1,043 |
| 1986 | 212 (22) | 132 (14) | 620 (64) | 964 |
| 1987 | 243 (25) | 115 (12) | 619 (63) | 977 |
| 1988 | 355 (30) | 129 (11) | 696 (59) | 1,180 |

*Source*: Yoon (1990: 111, table 4)

*Note*: figures in parentheses represent percentage of total investment

The private sector also outperformed the state in R&D activities. The explosive expansion of R&D in Korea in the 1980s was initiated by the private sector. In the 1970s, as Table 8.4. shows, the government was the major source of R&D activities in Korea. But it was reversed in the 1980s as a result of vigorous R&D expansion by the *chaebol*, mainly in the electronics industry. The state sold most R&D facilities of KIET to private firms as 'three largest semiconductor firms had far superior R and D resources to those of KIET'.[119]

Even when the Korean semiconductor industry faced a serious recession in the mid-1980s just after its ambitious investment had been commenced,[120] the Korean state did not provide any specific supportive measure except those already carried out. Nor did the *chaebol* demand direct intervention from the state.[121] Instead, the *chaebol* devised different corporate strategies for survival according to their different degrees of commitment to the semiconductor business and their technological levels.

Samsung, the head-starter which was most strongly committed to the semiconductor business, accelerated its catching-up effort by producing 256K DRAM earlier than initially planned and by diversifying its semiconductor products.[122] Hyundai did not reduce its investments, either, but it reorganised its production facilities by closing down its pilot operations in California and 'opted to become a foundry for other companies [in particular for TI]' while it was developing its own design through various technological agreements.[123] Goldstar also moved to foundry business mainly for Hitachi. It began to increase investment in DRAM production significantly from 1989 after ending its relation with AT&T.[124] Daewoo cancelled a large part of its initial investment plan and began specialising in customised chips by acquiring Zymos, a small US ASIC design firm, in 1986.[125]

*Table 8.4* Major R&D indicators in Korea (billion, Won, %)

| | *1976* | *1981* | *1986* | *1988* | *1990* |
|---|---|---|---|---|---|
| R&D expenditures | 60.90 | 293.13 | 1,606.9 | 2,454.1 | 3,349.8 |
| Funds from government | 39.18 | 121.73 | 374.3 | 522.9 | 651.0 |
| Funds from private sources | 21.72 | 171.40 | 1,232.5 | 1,931.2 | 2,698.8 |
| Gov't: private (%) | 64:36 | 42:58 | 23:77 | 20:80 | 19:81 |
| R&D/sales (%) | 0.36 | 0.67 | n.a. | 1.61 | n.a. |
| R&D/GNP (%) | 0.44 | 0.65 | 1.77 | 1.94 | 1.95 |

*Source*: adapted from L. Kim (1993: table 11.3); MOST (1993: tables 1.2 and 2.1)

### *Pattern of substitutions in comparison with the Japanese case*

According to Gerschenkron's schema, the state is the substituting institution for an extremely backward country to compete with its forerunners. In the iron and steel industry, we observed that his schema broadly fitted the comparison between Japan and Korea. The Korean state certainly substituted for the underdevelopment of its private sector in establishing and sustaining POSCO (section 7.2). In the catching-up in the semiconductor industry, however, the Korean state played only a limited role. Until the 1970s, its role was constrained mainly because of the lack of its capability to intervene in the process of technology transfer. In the 1980s, the private sector, mainly the *chaebol*, played a more significant role, as we discussed above. As far as the role of the state is concerned, it is difficult to find the evidence that the Korean state played any significant compensating role in competing with Japan. The collaborative research from the development of 4M DRAM might be a possible candidate in this regard. But it is still contentious whether this collaborative research played a decisive role in developing new generations of DRAMs as the Japanese VLSI Project had done.[126] How, then, can we explain Korea's catching-up with Japan in the semiconductor industry?

It seems to us that the answer can be attained by probing *substitutes* employed by Korea as compared to Japan. If we compare the structure of the Korean semiconductor industry with that of the Japanese, it is evident that the former is more *unbalanced*. This different degree of imbalance has to do with the different degree of backwardness.

First, the Korean semiconductor industry is characterised by a much weaker backward linkage, importing most semiconductor manufacturing equipment. The Japanese producers could reduce the ratio of equipment import from 70–80 per cent in 1976 to around 50 per cent by 1980 with the VLSI Project.[127] But Korea imported 97 per cent of the equipment and 90 per cent of raw materials from abroad, even in 1989 when it firmly established itself as a major producer of DRAM in the world market.[128]

This was surely a reflection of relative backwardness of Korea as compared to Japan when it began its catching-up process. Japan could embark on catching-up in equipment manufacturing and raw materials almost parallel with that in DRAM manufacturing mainly because it already had a relatively developed capital goods sector. But Korea could not proceed with catching-up in both manufacturing semiconductors and equipment manufacturing at the same time. Just as it had to wait for its catching-up in the semiconductor industry until its consumer electronics industry was strengthened during the 1960s and 1970s, its semiconductor equipment manufacturing sector could not develop until the semiconductor manufacturing sector was sufficiently developed. This is a typical pattern of Hirschman's import-substitution through the backward linkage effect in developing countries.[129] But, if Gerschenkron's terminology is employed, Korea's lack of a capital goods sector was partly substituted for by importing equipment.

Second, the Korean semiconductor industry had a much weaker forward linkage with its domestic market. The Japanese semiconductor industry developed initially based on its domestic market, as we pointed out before (section 8.2). Then, as its volume of production was increased, the share of export to total production was increased gradually.[130] The Korean semiconductor industry had virtually no relation with its domestic market. As Table 8.1 shows, most semiconductors produced were exported even in the 1980s. Until the 1970s, the export ratio was high because Korea was just an assembly point for MNCs. But the ratio remained still high, even after indigenous wafer manufactures developed during the 1980s, because the Korean wafer manufacturers were also highly export-oriented.[131]

This high degree of export orientation can be again attributed to the relative backwardness of Korea as its forward linkage industries such as computers and telecommunications did not develop enough to provide the semiconductor industry with sufficient demand pull. Moreover, even if the Korean computer and telecommunications industries had developed significantly, Korea's domestic market would have been too small to realise benefits of scale economy. In order to produce competitive products, it was essential to exploit the economies of scale on the global level, as in the case of the iron and steel industry.[132]

Therefore, even in the 1980s when the Korean computer and telecommunication industries were growing rapidly mainly through the government's promotional measures, they were not closely related with the domestic semiconductor industry. Most high-performance and complex semiconductors needed in computers and telecommunication equipment were imported.[133] The Korean semiconductor manufacturers concentrated their efforts on exporting standardised products, mostly DRAMs. Thus, the Korean strategy in the semiconductor industry even in the 1980s can be still understood as EP-*cum*-IS as compared to Japan's IS-*then*-EP. The export market was thus partly substituted in the Korean semiconductor industry for the role played by the domestic market in the Japanese semiconductor industry.

Third, the lack of technologies in Korea was compensated for by much wider foreign sourcing. In the earlier period of the Korean semiconductor industry, it was partly compensated for by FDI. As the continual relaxations of government's regulation on technology inflow indicates, the dependence on foreign technology was much greater in Korea than in Japan.

One significant factor in acquiring advanced technologies was the establishment of the outposts at Silicon Valley, mostly by acquiring small ailing US firms. Technologies owned by these firms became a significant source of advanced technologies for the Korean producers.[134] Moreover, Korean-Americans recruited for operating those outposts contributed decisively to the initial technological development.[135] Japanese semiconductor manufacturers actually pioneered establishing outposts in Silicon Valley from around 1978 by acquiring small American semiconductor firms,[136] but their functions seem to have been confined mainly to monitoring technological trend and acquiring specific technologies, since the main R&D activities were undertaken in Japan. In the Korean case, by contrast, the most crucial researches on semiconductor design were undertaken in the Californian outposts in its initial stage of catching-up.

Those substitutions mentioned above would be, at best, a partial compensation for the lack of technologies or domestic linkages in the Korean semiconductor industry. They themselves are not sufficient to answer the question as to how Korea could attain a comparable level of competitiveness to that of Japan. We suggest that it has to do with intensive concentration of investment and technological activities on narrowly defined products and technologies, i.e., *manufacturing of DRAM backed by the* chaebol *structure*.

From the beginning of their ambitious investment in the early 1980s, the Korean producers were well aware of the need to narrow their target. In the semiconductor industry, as emphasised before (sections 8.1 and 8.2), it was imperative for the latecomer to adopt an 'overtaking' strategy at some stage, because state-of-the-art technologies change so rapidly and therefore the risk involved in large-scale investment is too great. In order to overtake forerunners with limited resources, it was necessary for the latecomer to focus its effort on a narrowly defined area. This was the major reason why Japan focused its attention on DRAM in the VLSI Project, as discussed before.

The Korean producers concentrated their catching-up effort on DRAM for the same reason.[137] But they concentrated on DRAM far *more* than their

Japanese counterparts. In 1990, the share of DRAM in total semiconductor production of Samsung and Hyundai was 67.0 per cent and 77.6 per cent respectively.[138] In Japan, the degree of specialisation in memories or DRAM was not so high as that in Korea, although it was much higher than that in the US, since the forward linkage of the Japanese semiconductor industry was much stronger than that of Korea. For instance, the ratio of memories to total semiconductor production was only 26.9 per cent in 1980 when Japan began overtaking the USA in production of memories.[139] Korea's concentration on DRAM production became possible only by relying on the export market.

Moreover, the technological effort of Korea was concentrated more on *manufacturing* technologies. As we pointed out above, Korea imported most of the raw materials and equipment until it reached the forefront of the DRAM manufacturing. Korea narrowly focused its technological effort on producing internationally competitive DRAMs as soon as possible. What seems important to note here is the emphasis on 'learning-by-doing'. Although the Korean semiconductor producers increased their R&D investment significantly all through the 1980s, most of their investment went to facilities, as Table 8.5 shows. The ratio of investment in R&D to total investment was gradually increased only after investment in facilities progressed substantially.

Archambault's (1991) compilation of fragmented evidence is suggestive in this regard. In the case of Samsung, its entry into the production of 64K DRAM was 'quite late in their life cycle'. But the rationale for the late entry

*Table 8.5* Investment trend of Korean wafer manufacturers ($ million, %)

| | *1984* | *1985* | *1986* | *1987* |
|---|---|---|---|---|
| *Total investment* | 586 | 528 | 215 | 205 |
| *Facilities* | 544 (92.8) | 478 (90.5) | 179 (83.2) | 159 (77.5) |
| Samsung | 283 (94.6) | 268 (90.2) | 113 (80.7) | 106 (79.1) |
| Hyundai | 131 (94.2) | 129 (91.4) | 19 (90.4) | 22 (73.3) |
| Goldstar | 96 (85.7) | 61 (89.7) | 30 (88.2) | 11 (73.3) |
| KEC | 34 (91.8) | 20 (90.9) | 17 (85.0) | 20 (76.9) |
| *R&D* | 42 | 50 | 36 | 46 |
| Samsung | 16 | 29 | 27 | 28 |
| Hyundai | 7 | 12 | 2 | 8 |
| Goldstar | 16 | 7 | 4 | 4 |
| KEC | 3 | 2 | 3 | 6 |

*Source*: Choi (1991: 28, tables 2–12)

> was to gain the production and marketing experience that would enable Samsung to develop rapidly and then to manufacture 256Kb RAM chips, the following generation of memory ICs. The experience gained in producing these 'close to the technological frontier' 64Kb chips would also be used to produce a variety of other ICs: EEPROM, microprocessors and other memory chips with greater money-making potential.[140]

The emphasis on learning-by-doing was more pronounced in Hyundai. Hyundai almost finished building a plant for producing 64K DRAM in 1985 even without having the necessary R&D facilities and with a minimum number of qualified engineers. It did not have 'patience' 'to take time to recruit and train technical staff'.[141] After the initial period of trial and error, Hyundai refocused its semiconductor production on foundry work, a business primarily premised on manufacturing facilities. The vice-president of R&D for Hyundai Electronics explicitly stated that '[f]oundry work is more desirable because it helps you to stabilize your manufacturing technology much more quickly'.[142] In the beginning, 80 per cent of Hyundai's production was foundry work. But the ratio was rapidly reduced to 50 per cent in 1988 as Hyundai was accumulating its own technologies through learning-by-doing and by acquiring designs from other small US companies. Goldstar focused on foundry work for the same reason when it significantly increased its investment in DRAM production from 1989.

This *unbalanced* emphasis on DRAM manufacturing was sustained by the strategic concerns of the *chaebol*. In the previous section, we emphasised the synergy of the Japanese integrated electronics firms which came from technological and demand-side linkages, and financial support. The Korean semiconductor producers are mostly integrated electronics manufacturers, as are their Japanese counterparts. But the technological and demand linkages within the Korean firms were much weaker than in Japan, as we pointed out above. Instead, the Korean semiconductor producers could sustain their catching-up effort with the full support from their *chaebol* structure, and this support was mainly directed at financing the prolonged investment.

This difference has to do with the difference in ownership structure. Whereas the Japanese *keiretsu* is a loose federation without its own central office, the Korean *chaebol* still retains a centralised coordination system whose power is concentrated at the top, mainly to the owner of the group.[143] Once a strategic decision was made, therefore, the Korean *chaebol* could mobilise and concentrate its resources on the strategic sectors with more determination and persistence. This seems to be a major reason why Korean producers, some with existing electronics business much smaller and less developed than the Japanese or others with no prior experience in the electronics at all, could face direct competition with forerunners without receiving significant support from the state.

The size of leading Korean *chaebols* like Samsung or Hyundai is in fact not much smaller than the Japanese vertical *keiretsu* like Hitachi or Toshiba, though they are much smaller than the Japanese horizontal *keiretsu*.[144] By concentrating their financial strength on the limited segment of the semiconductor industry, Korean *chaebols* were able to sustain the prolonged race of investment.[145] Owing to the strategic concern of the *chaebol*, Korean wafer manufacturers could sustain an investment to sale ratio of over 50 per cent in the 1980s.[146]

This sustained investment bore fruit especially in the cases of 1M DRAM and 4M DRAM. While Japanese producers invested conservatively, expecting a slow growth of market demand, Korean producers invested heavily in facilities for 1M DRAM and 4M DRAM simultaneously. In 1989, Korea's facility investment amounted to 20 per cent of total investment in the world semiconductor industry.[147] As a result of this ambitious investment, Samsung became the largest producer of 1M DRAM in 1991 and 4M DRAM in 1993.

## 8.4 CONCLUDING OBSERVATIONS

The case of the semiconductor industry discussed so far reveals a very significant limitation of Gerschenkron's original schema in which two different latecomers are supposed to take basically the same catching-up strategy although their institutional patterns are different. In other words, it shows the existence of 'threshold' in adopting a Gerschenkronian catching-up strategy. Before reaching this threshold level, Korea had to take a roundabout route by pursuing import-substitution in the consumer electronics and by engaging in assembly of semiconductors for nearly two decades. In a sense, Hirschman-type gradual catching-up through backward linkages seemed necessary before Korea could embark on its ambitious catching-up in the 1980s.

This difference came mainly from different mechanisms of technology transfer between the iron and steel industry and the semiconductor industry. In the former, the pace of technological progress was rather slow and it was mostly embodied in capital equipment. Therefore, if capital for the initial large-scale investment was secured, even an extremely backward country could embark on a rapid catching-up to the forefront. But, in the latter, the pace of product and process innovation was very rapid, and it was difficult for latecomers to secure and assimilate state-of-the-art technologies. Thus, Korea could not follow the Japanese catching-up strategy from the beginning. It could employ only very limited substitutes, such as allowing FDI, pursuing import-substitution in consumer electronics, and so on.

Because of the limited degree of substitution in Korea, the role of the state was different from Gerschenkron's original schema: the state intervention was more pervasive in a moderately backward country, i.e.,

Japan than in an extremely backward country, i.e., Korea. In terms of the degree of the state intervention in the overall economy, the Korean state has been certainly more interventionist than the Japanese state during the post-war period as it was in the iron and steel industry. But the semiconductor industry was an exception. The Korean state simply did not have capability to intervene. Allowing freer play of market by MNCs and technology inflows was a better option for the Korean state to take if it wanted to enter the industry. And, in the process of the technological catching-up in the 1980s, the structure of the *chaebol* mainly compensated for Korea's relative backwardness with Japan.

However, our comparison of the Japanese and Korean catching-up in the semiconductor industry conforms to Gerschenkron's pattern in its functional aspect, that is, latecomers pursued a more unbalanced growth. Japan initially concentrated on DRAM manufacturing and this unbalanced emphasis on DRAM manufacturing was much more pronounced in Korea. In this respect, the catching-up processes of Japan and Korea in the semiconductor industry were microcosms of Gerschenkron's pattern at the level of an industry.

According to Gerschenkron's observation, latecomers in the latter half of the nineteenth century like Germany and Russia concentrated their catching-up effort on the producers' goods industries because the overall technological trend was then towards increasing capital intensity. In the catching-up in the semiconductor industry during the latter half of the twentieth century, the technological trend was rather different, moving towards increasing both capital and R&D intensity. Here, unbalanced emphasis on DRAM manufacturing was promising to Japan and Korea in reaching the forefront as fast as possible.

The unbalanced focus of the catching-up effort on a narrow segment of the semiconductor industry was possible only by utilising the international market. But we pointed out that the degree of the utilisation of the international market in the initial period of catching-up was related to the degree of backwardness. Japan, as a moderately backward country, could adopt IS-*then*-EP, basing its initial catching-up on its domestic market. But this option was not available to Korea because its backward and forward linkage sectors did not develop. It had to rely on the export market from the beginning. Thus, we described the Korean strategy as EP-*cum*-IS.

# CONCLUSIONS

Discussing Weber's interpretative sociology (section 3.1.1), we emphasised that his sociology was basically a tool for understanding the historical process. The essence of Weber's analysis does not lie in the tools themselves, but in their elaboration and extension in explaining the actual historical process by which our understanding of social phenomena is progressed. Gerschenkron's 'pattern of industrialisation' was proposed in the same spirit. It is an ideal type for understanding the process of catching-up. As it is at an intermediate level of abstraction, however, it has its own temporal and spatial limits, being conditioned by the technological trend and international rivalry of Western Europe during the latter half of the nineteenth century. We attempted to extend Gerschenkron's schema into different periods and countries. This was not directed at refutation or confirmation of his schema. Rather, its aim was to enlarge our understanding of the process of catching-up by elaborating the schema in extended horizons and elucidating the variations from the original pattern. Conclusions emerging from our attempt can be summarised as follows.

## 1 GERSCHENKRON VS. HIRSCHMAN: SUBSTITUTE VS. THRESHOLD

In comparing 'late-late' industrialisation in the twentieth century with 'late' industrialisation in the nineteenth century, Hirschman (1968: 94–5) argues that, among Gerschenkron's six propositions about patterns of industrialisation (section 3.2.1), 'only the last one [which is concerned with the role of the demand from the agricultural sector] applies unconditionally to the late late industrializers'.[1] The role of 'special institutional factors' (point 5), he contends, 'became important in most of Latin America *after* the ISI [import-substituting industrialisation] had already been underway as a result of private, decentralized initiative for a considerable time'.[2] And, as to the remaining four points, he argues that

> almost the opposite could be said to hold for our late latecomers. Their industrialization started with relatively small plants administering 'last touches' to a host of imported inputs, concentrated on consumer rather than producer goods . . . Even though the rates at which new plants were built and at which their output expanded were often respectable, the process . . . lacked some of the essential characteristics of Gerschenkron's 'great spurt'.
>
> (Hirschman 1968: 94–5)

Hirschman attributes these differences mainly to the different *size* of the technology gap the late-latecomers face. He stresses that the latecomers in Western Europe in the nineteenth century 'were already producing their own capital goods, if only by artisan methods' and were able 'to jump with both feet into a newly emerging dynamic industrial sector' by designing and often manufacturing special equipment required.[3] For Hirschman, the latecomers in the nineteenth century could commence their catching-up based on the producers' goods sector, not only because technological progress was 'more vehement' and thereby the catching-up prospect was brighter in this sector, as Gerschenkron argues, but also because they had a capital goods sector already developed to some extent. However, the late-latecomers could not commence catching-up in the producers' goods sector from the beginning because they hardly had a capital goods sector and therefore faced a much bigger technological gap with advanced countries than the latecomers had done. Therefore, according to Hirschman, late-late industrialisation became 'a *highly sequential*, or *tightly staged*' process of ISI, starting 'predominantly with the manufacture of finished consumer goods that were previously imported and then moves on . . . to the "higher stages" of manufacture, that is, to intermediate goods and machinery, through backward linkage effects'.[4]

Hirschman is quite correct in pointing out the possible existence of a threshold level in commencing a Gerschenkronian catching-up strategy. In our comparison of catching-up between Japan and Korea in the semiconductor industry, we could detect the existence of a threshold. Only in Japan, a moderately backward country, could the ambitious Gerschenkronian style catching-up strategy be adopted (section 8.2). For an extremely backward country like Korea, which faced too large a technology gap with the American MNCs, it was necessary to adopt the Hirschman-type gradual ISI process of catching-up starting from assembly production and moving on to more technologically advanced processes mainly through the backward-linkage effect. The Gerschenkronian type catching-up process was undertaken in the 1980s only after its electronics industry developed and *chaebols* grew to some extent (section 8.3).

However, Hirschman fails to pay attention to the fact that different patterns of catching-up also come from different *characteristics* of the technology gap. As emphasised in section 3.2.1, the Gerschenkronian strategy of establishing 'bigger and bigger' plants with latest technologies was effective when the pace of product and process innovation was not so rapid, when its direction was towards increasing economies of scale, and when technological progress was mostly embodied in capital equipment. This strategy was applicable to a late-latecomer, i.e., Korea, in the iron and steel industry because the same technological trend still persisted (section 7.2).

The gradual ISI process of Korea in the semiconductor industry as compared to that in the iron and steel industry was also affected by different characteristics of the technology gap. The rapid pace of product and process innovation in the industry increased the risk of initial large-scale investment on the part of the latecomer. But, since the production process of semiconductors was spatially separable, MNCs embarked on the global production systems by setting up their assembly points in developing countries. Korea and other East Asian countries, in turn, were able to enter the industry without risking large-scale investment on facilities and R&D from the beginning.

The development of the iron and steel industry in Prussia during the first half of the nineteenth century showed that the ISI strategy was not exclusively related to the size of the technology gap. Here, contrary to Hirschman's view, a gradual process of the ISI strategy was taken in a moderately backward area, i.e., in Rhineland–Westphalia, not in an extremely backward area, i.e., in Upper Silesia. This was because the combination of old-style labour-intensive process in a later stage of the iron and steel production process with imported pig-iron was an effective way for catching-up in Rhineland–Westphalia since the iron- and steel-making process was not so integrated during this period (section 5.2). And the absence of a capital goods sector in Upper Silesia could be substituted for by introducing foreign engineers (section 5.1).

Our study did not provide an exhaustive comparison of late vs. late-late industrialisation mainly because it focused only on two industries, the iron and steel industry and the semiconductor industry. We did not probe the patterns of catching-up at the level of the economy, which was the main concern of Gerschenkron's and Hirschman's analyses. By concentrating on the industry-level evidence, however, we could provide room for accommodating both Gerschenkron's and Hirschman's observations while delineating their respective limitations.

Gerschenkron does not give much importance to the existence of the threshold level for catching-up though he acknowledges it in the textile and machinery industries rather as exceptional cases (section 3.2.1). This is mainly because he views late industrialisation as a process of employing *substitutes*. His schema is therefore constructed with substituting institutions and strategies. Gerschenkron, however, does not directly tackle the question of difficulty in employing substitutes. On the other hand,

Hirschman puts a special emphasis on the existence of the threshold levels faced by the late-latecomers as compared to the European latecomers in the nineteenth century. His ISI strategy can be understood as one taken by latecomers who have difficulty in employing substitutes. Our study shows that the threshold level played an important role in the catching-up in the semiconductor industry whereas it was not so significant in the iron and steel industry. Further research dealing with other industries would provide us with a comprehensive picture of the patterns of catching-up at the level of the economy.

## 2 GLOBALISATION AND EXPORT-ORIENTED INDUSTRIALISATION

In proposing ISI strategy for the late-latecomers, Hirschman also neglects the possibility of industrialisation through export specialisation in the twentieth century. He seems to assume that the late-latecomers can export industrial goods only after import-substitution has progressed to some extent. This would be mainly because his conception of the developmental process is based on the experience of Latin American countries which were initially specialised in agriculture in the world economy. Here, a main impetus for ISI was 'growth of domestic market (as a result of [agricultural] export growth)'.[5] Focusing on this domestic demand linkage in Latin American countries, Hirschman seems to assume that industrialisation begins only with the production of *finished goods consumed in the domestic market*. He emphasises 'the special difficulty of moving into export markets' in this respect.[6]

However, the development of the semiconductor industry and some segments of the electronics industry, which were not confined to finished goods, in Korea and other East Asian countries during the post-war period began by relying almost exclusively on the export market (section 8.3). This was because the increased globalisation, especially in the electronics industry, during the latter half of the twentieth century opened the possibility of catching-up through participating in the export market from the beginning. When East Asian countries adopted their export-oriented industrialisation (henceforth EOI) strategy, they did not seem to have clearly envisaged its inherent advantages. Mainly due to lack of natural resources and an insufficient agricultural base, they were forced to export manufactured goods.[7] Assembly of semiconductors and other simple electronics products for the export market was mainly regarded as a means for earning foreign currency. But, in retrospect, this EOI strategy had several advantages over simple ISI strategy in facilitating technological progress in latecomers.

First of all, the export market not only provided stronger competitive pressure but also forced East Asian manufacturers to constantly upgrade their technological level. Since they entered the electronics industry mainly based on their low wages, it was inevitable for them to upgrade their technological level as their wage levels were increased in the course of development. And, in order to upgrade technological level in the electronics industry where technological change was rapid, it was imperative to introduce advanced technologies continuously by adopting a liberal policy on technology transfer (section 8.3). Second, since most electronics products were exported to the advanced countries, the continuous information flow from sophisticated users in these countries contributed to facilitating technological progress in the East Asian manufactures. In particular, the role of the US mass merchandisers in providing technical assistance was important in keeping up with changes in technologies and consumer preferences.[8] In a sense, the US OEM buyers substituted for the role of technical guidance which should have been assumed by the East Asian manufacturers themselves or their governments. Third, the export market enabled the East Asian manufacturers, especially those of Korea, to exploit economies of scale which could not be realised in their domestic markets. Mass production of consumer electronics goods like TV sets, VCRs, microwave ovens, and so on, and heavy investment in wafer manufacturing facilities by Korean electronics firms would have been impossible without relying on the export market.

These constant pressures for upgrading technological level, continuous inflow of advanced technologies and utilisation of economies of scale on the global level were often absent in the simple ISI process of many other developing countries. For instance, it is often pointed out that the Brazilian electronics industry still remains fragmented and uncompetitive mainly due to its reliance on the domestic market. According to Dahlman and Frischtak (1993: 427, table 13.8), only four firms among ninety R&D active electronics firms were engaged in export activities in the mid-1980s. Owing to this low level of export orientation, 'government policies and institutional support mechanisms [for R&D have] . . . been insufficient to close in the technological frontier'.[9] Moreover, the competitiveness of electronics producers in Latin American countries was further undermined by the adoption of the import-substituting policy in capital goods sectors relying only on the domestic market and the strict regulation of the process of technology transfer. The result was the emergence of 'an overly diversified capital goods industry lacking economies of scale or specialization necessary to reduce costs, keep up with new technologies'.[10]

In East Asian countries, however, advanced capital goods were imported freely in so far as they were employed in export activities.[11] And, as emphasised in our study of the Korean electronics industry, more liberal policies in technology inflow have been continually introduced as the problem of technological upgrading has become more acute over time. These policy measures were imperative in maintaining the competitiveness of their finished products in the international market. Viewed in this way, the relatively successful catching-up in the electronics industry of East Asian countries as compared to Latin American countries can be explained by the existence of the built-in mechanism for upgrading technological level in their EOI strategy.[12]

A comparison between the 'moderately backward' countries after the end of the Second World War, i.e., between Japan and West European countries in the semiconductor industry, points to the similar importance of an export-oriented strategy. Japan successfully caught up with the USA in the semiconductor industry from the late 1970s. Although its catching-up, especially in its initial period, was more based on domestic linkages if compared to its follower, Korea (sections 8.2 and 8.3), the overall direction of its catching-up process was geared at the export market.[13] But the European manufacturers were principally geared to the domestic market until the mid-1980s.[14] The result was continuous falling behind in the world market.[15] The European manufacturers could later narrow the technology gap with frontier producers only by adopting 'outward-looking global strategies' in conjunction with the recruitment of world class professional managers who had worked in the US firms, the acquisition of innovative US firms, and the establishment of Europe-wide R&D programmes and alliances.[16] Participation in global competition seems to have been an unavoidable option in catching-up in the electronics industry in general, and the semiconductor industry in particular.

On the other hand, the increased globalisation in the electronics industry seems to have opened a wider range of options for catching-up. In the iron and steel industry, it seems necessary for a latecomer to have taken the Gerschenkronian catching-up strategy, even in the twentieth century, if it wanted to be competitive in the international market. This was mainly because technological imperatives in the industry were still towards increasing economies of scale. In the electronics industry, however, latecomers could join the international production network in various ways. A comparison between Korea and Taiwan would be illustrative in this regard.

Both were 'extremely backward' countries as compared to Japan and the USA at the beginning of the post-war period. And the electronics industries of both countries started from simple assembly relying on their low wages. However, once their wage levels were increased in the course of industrialisation and thereby the need for international competition based on technologies became acute, they went for rather different routes for catching-up. Korea relied more on mass production of finished goods which required heavy investment in production facilities. This catching-up path was partly a way of utilising its *chaebol* structure, though it stood in more direct competition with its forerunners like Japan and the USA. Thus, a Gerschenkronian pattern emerged: Korea's catching-up was most pronounced in a capital-intensive segment, i.e., DRAM production (section 8.3).

Taiwan, by contrast, pursued a 'niche market' strategy utilising the flexibility of its small firms, relying more heavily on international subcontracting networks. And Taiwan concentrated more on component production which was 'the most labour-intensive part of the electronics industry' until the early 1980s than Korea did.[17] In the semiconductor industry, Taiwan has been stronger in ASICs, which is appropriate to small firms, while weaker in DRAMs which require large-scale investments if compared with Korea.[18] As shall be discussed below, these differences in catching-up strategy and product specialisation between the two countries were more to do with their differences in history and political economy than with their different perceptions of efficiencies inherent in those strategies.

## 3 UNBALANCED GROWTH

Most controversies over the 'balanced' vs. the 'unbalanced' growth strategies in the 1950s and 1960s proceeded from the viewpoint of the domestic economy of developing countries.[19] The balanced growth strategy was advocated on the basis of a supposition that economic growth is hampered if domestic supply factors such as infrastructures, skilled labour, backward linkage industries, and so on are constrained, or if domestic demand from the agricultural sector or forward industries cause bottlenecks. Hence the need for a 'big push' across all sectors. By contrast, the unbalanced growth strategy was proposed on the basis that a simultaneous investment across sectors is difficult to make in developing countries due to the lack of resources. Rather, it was argued that once investments in some selected sectors are made, bottlenecks caused by these investments would 'induce' further investment.

Even if only the domestic economy is considered, we are inclined to support the view of unbalanced growth, since balanced growth requires too many prerequisites at the same time. As Hirschman (1958: 53–4) pinpoints, 'if a country were ready to apply the doctrine of balanced growth, then it would not be underdeveloped in the first instance'. If the international aspects of economic development are considered, the plausibility of balanced growth is further weakened for the following reasons.

First, technology transfer itself is a creation of imbalance in developing countries. It is certainly true that foreign technologies are often adapted and modified to domestic resource conditions. A developing country may choose only those technologies which are appropriate for its domestic resource conditions.[20] However, initial domestic conditions should be also adapted to foreign technologies simply because they are 'new' to developing countries. Moreover, not all the advanced technologies are malleable.[21] Generally speaking, a more advanced technology tends to create a greater imbalance in a recipient country and is more likely to stimulate technological progress in latecomers through 'learning-by-doing' (sections 2.1 and 8.3). According to the theories of balanced growth, the dual economy frequently found in developing countries is regarded simply as a distortion. But, from the viewpoint of unbalanced growth, it can be regarded as a consequence of unbalanced growth derived from the fact that the utilisation of foreign technology is much more vigorous in the modern sector than in the traditional sector.[22]

Second, if international competition between forerunners and latecomers is considered, unbalanced growth is an effective way for latecomers to compete. As we noted before (section 3.2.1), the essence of Gerschenkronian strategy is catching-up by accentuating unbalanced growth. The logic behind this is simple. Latecomers cannot catch-up with forerunners in all the industrial sectors at the same time. It is more promising for latecomers to focus their catching-up effort on those sectors where they have advantages or at least fewer disadvantages compared to forerunners. Latecomers' limited resources can be better utilised if devoted more to winning games than to losing ones. Thus, the nineteenth-century latecomers concentrated their catching-up effort on the producers goods' sector, as Gerschenkron observes. And the twentieth-century latecomers in the semiconductor industry put an unbalanced focus on DRAM manufacturing, and this imbalance was more pronounced in an extremely backward country like Korea than in a moderately backward country like Japan (sections 8.2 and 8.3).

Third, unbalanced growth seems inevitable with globalisation. If a country participates in the international production network, its system unavoidably becomes unbalanced from the viewpoint of its national economy. Thus, in the development of the Korean economy, the balance between the agricultural and the non-agricultural sectors was entirely neglected. And within the semiconductor industry, this imbalance was also pronounced as it has developed without much significant linkage with its domestic electronics industry or capital goods sector (section 8.3).

## 4 THE ROLE OF INSTITUTIONS: MARKET VS. STATE

We emphasised that the efficiency of institutions cannot be assessed separately from their contexts (section 3.1). In an ideal world, one can construct an ideal type which shows the efficiency of either the market or the state. Thus, transaction cost economics 'has been invoked in support of both market pessimism and market optimism', as Matthews (1986: 907) points out. Neo-classical theory is a typical ideal type which shows the efficiency of the market. But if market failure is introduced in this model, it can be easily turned into an ideal type advocating state intervention. The question of relative efficiency between the market and the state cannot be resolved in the domain of ideal type. It can be assessed only by analysing concrete contexts in which those institutions function. Are there, then, any general 'lessons' regarding the relative role of the state and the market from our case studies?

In answering this question, it seems important to recognise the fact that the state is situated between the international market and the domestic market. It is almost impossible for a particular state, especially that of a latecomer, to dictate international market conditions or the technological trend. It has to conform to those external conditions. The scope for the role of the state is bound by external conditions. For instance, the Korean state had to take more liberal policies on technology transfer in the electronics industry because it was a better way to upgrade technological level in the given conditions of technological trend and technology market in the electronics industry (section 8.3).[23]

On the other hand, the state is at a higher level of institutional hierarchy within a country and it is always present, passively or actively, in its domestic economy. In the long-term process of late industrialisation, the role of the state as an institution of the higher order can be conceptually divided into two: an institution builder and an institution manager. After a long delay in the process of industrialisation, the state of a latecomer should establish new institutions to initiate catching-up. The historical significance of the Prussian Reform, the Meiji Restoration of 1868 in Japan, and the military coup of 1961 in Korea should not be missed in this regard. As one historian aptly puts it, '[h]istorically, the most important way in which the state stimulated industrial growth in a capitalist setting was through its ability to restructure the institutions of society – i.e. through its ability to *create* a capitalist setting in the first instance'.[24] Once new institutions were set up, they should be managed. The state has ultimate responsibility as well as power for this management. This is undertaken by enforcement of laws, regulations and various policy measures. In this respect, many scholars point out the importance of long-term political stability in successful industrialisation.[25]

The state's function as an institution builder and most of its functions as an institution manager would correspond to Gerschenkron's negative roles of the state. In this role, as emphasised before (section 3.2.3), the market cannot be compared with the state simply because it cannot be an alternative to the state. In a sense, this role of the state is always crucial in any country. Statist interpretations of economic history seem to stress the importance of the role of the state often by including the negative role.

We argued that this comparability problem would be one major reason why Gerschenkron distinguishes between the negative and positive roles of the state. In his comparative schema, only the positive role of the state is included because it can be compared with that of German investment banks and that of the British market. And, in his schema, the efficiency of the positive role of the state is determined by the degree of backwardness. In a sense, this is very similar to the 'market failure' argument for state intervention. That is, state intervention is needed in a more backward country where market failure is more prevalent. But Gerschenkron's schema is more than a market failure argument.

First, since it adopts a 'three country paradigm', it is more sophisticated than the market vs. state dichotomy. In particular, it explains different forms of coordination within the market, that is, the difference between hierarchical coordination within the market, i.e., coordination by the investment banks, and non-hierarchical coordination within the market.[26] Here, the possible failures in price mechanisms do not necessarily call for state intervention because there is room for the private sector to correct those failures.[27]

Second, state intervention is not simply a corrective measure for malfunctions of markets. Rather, it provides what the domestic market in an extremely backward country simply cannot perform even if it may function perfectly. For instance, domestic resource mobilisation in Russian industrialisation required 'coercive' methods which the market mechanism could not provide no matter how lucrative the market incentives provided. Likewise, our extension of Gerschenkron's schema to the case of the semiconductor industry shows that state intervention was a function more of 'catching-up failure' than of simple market failure. The Japanese state increased the degree of its intervention as the USA continually forged ahead with product and process innovations (section 8.2). Here, the Japanese state intervention provided what the Japanese market simply could not do, however well-functioning it was.

The main focus of this book regarding the relative efficiency between the market and the state was placed in the context derived from the degree and characteristics of the technology gap. Our elaboration and extension of Gerschenkron's schema shows, more than anything else, that the utility of state intervention is not a constant variable. It changes with the degree of backwardness as the three countries with different degree of backwardness show different institutional responses. The utility of state intervention is also conditioned by characteristics of the technology gap. The state is bound by this external force. Thus, the Korean state has intervened in its semiconductor industry less than its Japanese counterpart has done (sections 8.2 and 8.3). The state intervention in its domestic market is also bound by political economy since it incurs its own costs, economically or politically. Here, we need to enter the area of the political economy of state intervention.

## 5 TECHNOLOGY AND POLITICAL ECONOMY

We emphasised that the schema adopted in this book is an ideal type which deliberately accentuates *economic* aspects of the catching-up process, leaving other aspects outside the schema (section 3.2.3). In our comparative setting, the role of political economy was not so pronounced because our sample countries, Germany, Japan and Korea, were all successful in making institutional transitions demanded by the imperatives for catching-up. But many unsuccessful attempts at catching-up can be explained in terms of political economy. Different routes for successful catching-up can be also understood by paying attention to different internal political conditions. In the actual process of history, technological and political aspects are often inseparably intertwined in determining the efficiency of institutions. The economic explanations proposed here should be complemented by political explanations. The real analytical problem would be how to construct an ideal type or a combination of ideal types for this purpose and how to explain those configurations in concrete cases. The following cases indicate the difficulty involved in the comparative analysis of such configurations.

If the development of the iron and steel industry of Rhineland–Westphalia is compared to that of France in the first half of the nineteenth century, the difference between the liberal and differential tariff system of Prussia and the prohibitive and monolithic tariff system of France is easily detected.[28] In fact, French iron and steel production remained almost stationary in the 1850s whereas that of Germany surged.[29] And it is often argued that 'the [French] tariff acted *as a premium on old fashioned methods of production*'.[30] The slower pace of the French catching-up was certainly related to the 'strait-jacket' of the French tariff system.[31]

However, once one considers the political economies behind those different tariff systems of France and Prussia, the assessment of their relative efficiency becomes problematic. It would require further research to answer the question as to why the liberal and differential tariff system was not resisted seriously in Prussia.[32] But the French protective tariff system was definitely a consequence of the Restoration politics. The French Revolution, which 'ground up the tissues and sinews of the old order until nothing remained but the individual particles',[33] brought about extreme individualism of small unit enterprises. And the government of the Restoration, which was paranoid about social instability, could not resist the demand 'from manufacturer to workman' for maintaining the protective tariff system.[34] Considering this situation, one might argue that the Prussian liberal and differential tariff system was not an option available to France during the period. Within the framework of the French political economy, the possible efficiency gains from adopting the liberal tariff system might have been more than offset by the political costs of social instability. Our discussion again leads to the conclusion that it is at best very difficult to make general propositions about the efficiency of institutions or policies across countries (section 3.1.2). Efficiency implications of policies or institutions can be drawn only from a concrete analysis of historical configurations of often conflicting factors in a given country.[35]

A similar argument can be made with regard to the comparison between EOI of East Asian countries and ISI of Latin American countries. It was pointed out above that the relative retardation of the Brazilian electronics industry can be explained by its failure in participating in the export market. But political economists would raise the following question: Was it because the Brazilian state could not foresee the advantages inherent in the EOI strategy or because it simply could not move to that strategy? It is often pointed out that Korea and Taiwan experienced only a short ISI period during the 1950s and soon moved on to EOI because there were 'no vested interests blocking structural change' or 'no entrenched bourgeoisie class'.[36] They were also 'forced' to export owing to their external conditions and resource endowment, as mentioned above. In Latin American countries, however, there had already emerged a strong group of industrialists whose interests were embedded in ISI policies, and related social institutions had been securely established through the long period of ISI process dated from the 1930s. Thus, one could argue that 'the much advertised noncompetitiveness of Latin American industry may be rooted more in the failure to modify institutions than in any inability to bring down real costs' and the reason 'has to do . . . with the distribution of power in Latin American societies', as Hirschman (1968: 119) advances.

On the other hand, differences of political economy between Korea and Taiwan affected only their patterns of catching-up, not their success or failure. Here, some patterns of substitution can be established. An important difference in political economy between the two countries is that Taiwan was a heterogeneous society which was divided between the mainlanders and native Taiwanese, whereas Korea was a homogeneous society. In Taiwan, the mainlanders retained political power over native Taiwanese by repressive measures. And the Taiwanese state restrained the growth of big businesses which might politically challenge the mainlanders' domination. Hence the emergence of a dual structure in the Taiwanese economy: the dominance of public enterprises in most strategic heavy and chemical industries and the dominance of small firms in the private sector.[37] By contrast, the Korean government promoted big businesses without any fear of political conflicts between the state and the private business sector.

Evidently, the small size of Taiwanese firms imposed some structural weaknesses on the economy. In particular, it had difficulty in large-scale resource mobilisation and in taking risks in long-term projects. These disadvantages were, however, countered by two substituting measures. One was the adoption of 'niche market' strategy mentioned above. Taiwan's private firms tried to penetrate those sectors where small firms are competent, mainly by exploiting the subcontracting network in the international market. The other substituting measure was state intervention. In the overall economic management, the state's direct intervention was more pervasive in Korea than in Taiwan.[38] However, once the Korean *chaebols* were firmly established in the 1970s, the Taiwanese state became more interventionist than its Korean counterpart, especially in relation to large-scale projects with high risks, because it had to substitute for the lack of large firms in its economy. Reflecting this reversed trend, the public sector contribution to the total R&D in the economy was 60 per cent in 1986 in Taiwan whereas it was around 20 per cent in Korea.[39]

The semiconductor industry is a representative case. We emphasised that the take-off of the Korean semiconductor industry in the 1980s was led by *chaebols* (section 8.3). But, in Taiwan, the existing electronics firms such as Tatung and Sampo 'opted not to enter the semiconductor market' in the 1970s.[40] Instead, a public institution, the Electronics Research and Service Organization (ERSO), has kept assuming the leading role in the Taiwanese semiconductor industry.[41] Taiwan's major semiconductor manufacturers in the 1980s were all 'spin-offs' of ERSO and they still rely heavily on ERSO as a source of new technologies.[42] It would be a subject of future research as to whether this kind of institutional substitution in the Taiwanese economy as compared to the Korean case has been, and will be, a perfect substitution.[43] But the fact remains that the Taiwanese electronics industry has been one of the most successful cases during the post-war period.

## 6 PLAUSIBLE EXPLANATIONS

The focus of this book was placed on attempting an interpretation of the catching-up process rather than digging out new facts from the vast pool of history. We emphasised the limitations of universal propositions in interpreting history. Some concepts employed in advancing universal propositions, such as social (or technological) capability or transaction cost, were found to be too vague to give meaningful understanding of history. And, if those concepts are reduced to some concrete factors and treated as 'prerequisites', another serious problem arises. For those supposed prerequisites are often substituted for by other factors in different temporal and spatial settings (sections 1.3 and 3.2).

Criticising the 'absolutism' of universal propositions, we took a relativistic position, following Gerschenkron and Weber, in presenting our interpretation (Chapter 3). The interpretative schema adopted here was that of intermediate level of abstraction. From the beginning, it aims at providing only a *partial* picture of the reality. And it was constructed as a differentiated system in order to accommodate temporal and spatial characteristics because we regarded the diversity resulting from those characteristics as critical in understanding the process of catching-up.

At the same time, we pointed out the limitations inherent in our explanatory schema. That is, first, its explanatory power is constrained by its own spatial and temporal limits, and second, it is focused primarily on 'economic' aspects of the catching-up process. As regards the former, we argued that the applicability of our schema has less to do with exact replication of its 'predictions' in different spatial and temporal settings. It has more to do with providing a 'tool' for comparative studies by which our historical understanding is facilitated. Although the original schema, i.e., the basic pattern, required some modifications and its scope for application was limited, its elaborations and extensions were the method adopted in our interpretation of the historical catching-up process.

As regards the latter limitation, we emphasised that the selection of a schema is never free from 'presuppositions'. It is 'conditioned by the orientation of our cognitive interest', not by any objectively predetermined criterion, as Weber (1949: 64) points out. There are multiple entries to interpreting the same phenomenon. One can start an analysis from either economic or political factors. There is no 'last analysis', as Gerschenkron (1968: 47) stresses. And we argued that non-economic factors which are not included in our schema should be understood by constructing a complementary ideal type, not by replacing our schema.

How plausible, then, is our explanation of the catching-up process? One major limitation of our case studies would lie in the fact that they did not directly confront other competing explanations about the same phenomenon. This could have been done only by constructing complementary ideal types and assessing the relative importance of different forces at particular historical configurations. But our study relied on only one ideal type, that is, the patterns of industrialisation derived from the degree of backwardness. We simply put forward our interpretation without attempting an 'ordering' of competing explanations in individual countries. For instance, in the cases of Japanese and Korean catching-up, many authors emphasise the importance of high levels of education in their successful industrialisations. Although we were critical about approaches dealing with the level of education as a prerequisite and emphasised the importance of 'learning-by-doing' (sections 1.1 and 2.1), we did not attempt to assess their relative importance in individual countries. Neither did we try to examine the weights between economic and non-economic factors in the context of particular countries, as pointed out above.

In this respect, our interpretation of the catching-up process did not pass a rigorous test. Therefore, McCloskey (1991: 100–1) argues that 'Gerschenkron's hypothesis about European industrialisation makes better sense if the language of "hypothesis" is dropped and that of *storytelling* substituted'.[44] As a story, it cannot be 'refuted' by other more plausible stories. It still remains as *one* story. And it can be enriched by other stories. Moreover, as pointed out above in discussing 'technology and political economy', it is often very difficult to establish exact and objective weights among interconnected factors. Landes (1969: 357) thus argues that 'when all is said and done, neither empirical evidence nor theoretical reasoning is likely to settle the dispute. Sharp differences of opinion will always remain.'[45]

The consequence is that we have ample stories, but meagre 'lessons'. What we have advanced in this book was no more than 'pointing at *potentially* relevant factors and at *potentially* significant combinations among them', which Gerschenkron (1962: 6) assumes as 'the historian's contribution'. We cannot provide 'four-lane highways' on which any latecomer can travel rapidly and safely. The configuration among technological trend, the degree of backwardness, catching-up strategies and institutional responses is diverse across countries and changes over time in actual history. The only general lesson that can be safely drawn from the past history would be 'that the policies toward the backward countries are unlikely to be successful if they ignore the basic *peculiarities* of economic backwardness', as Gerschenkron (1962: 30) points out.

# NOTES

## INTRODUCTION

1 Freeman (1989: 86) reminds us that Fredrich List, a principal architect of the German catching-up strategy, 'was so overwhelmed by the technological and economic leadership of Britain in the first industrial revolution that he died believing that Germany would never overtake Britain'. (See also Henderson 1975: 176.)
2 Rosovsky (1972: 238).
3 Khan (1989); Datta-Chauduri (1981).
4 A similar view of the interaction between forerunners and latecomers was advanced by Schumpeter (1934; 1939; 1943).
5 A similar view has been taken in explaining the long waves in the world economy. For Schumpeter (1939) and Freeman, Clark and Soete (1982), the 'swarming' process, rather than initial innovation, has the principal expansionary effect on the economy.

## 1 LATE INDUSTRIALISATION AND CATCHING-UP

1 Durlauf and Johnson, quoted in Broadberry (1992: 13).
2 In 1989, China's population was 1,113.9 million whereas the total of low-income countries is 2,948.4 million (*World Bank* 1991: table 1).
3 India's growth rate is 1.8 per cent, and that of other low-income countries excluding China and India is 1.4 per cent (*World Bank* 1991: table 1).
4 Abramovitz (1979; 1986); Baumol (1986); Dowrick (1991); Dowrick and Gemmel (1991); Maddison (1982, 1991); Broadberry (1992); Barro (1991); Verspagen (1993).
5 Abramovitz uses productivity and technology gap interchangeably, with some qualifications. Although this usage is not rigorous, we shall not distinguish between them in so far as our discussion is not severely constrained. For a more detailed discussion on the relation between productivity and technology gap, see Broadberry (1993).
6 Abramovitz (1986: 221). The page reference is to Abramovitz (1989).
7 Abramovitz (1986: 222–3). We shall discuss the conceptual robustness of social capability later in this section and in sections 2.1 and 2.2.
8 Abramovitz (1986: 222). It seems that, schematically, four different types of latecomers can emerge from different combinations of the technology gap and the social capability gap: 1) a country with a big technology gap but with a high level of social capability has the highest total potential for rapid catching-up, 2) a

country with a big technology gap and with a low social capability has the lowest total potential for catching-up, 3) a country with a small technology gap and with a high social capability may be on the verge of overtaking the forerunner, and 4) a country with a small technology gap and with a low social capability may be an *ex*-forerunner who has just begun falling behind owing to technological and institutional inertia in its society.

9 Abramovitz (1986: 229).

10 Abramovitz (1986: 225).

11 Although Abramovitz (1986: 230–9) proposes another qualified version in the case of the endogeneity of social capability, the foregoing seems sufficient to show his methodological approach to the phenomenon of catching-up.

12 Abramovitz (1986: 229). See also the following remark:

> For decades it [catching-up] operated only erratically and with weakened forces. The trouble at first lay in deficient social capability, a sluggish adaptation of education and of industrial and financial organization to the requirements of modern large-scale technology.
>
> (1986: 230)

13 Abramovitz (1986: 228, fn. 14) refers to Baumol (1986), who proposes this view, in a footnote after qualifying the bias of his sample.

14 Abramovitz (1986: 240) himself emphasises that his analysis shows 'only a tendency that emerges in *the average experience* of group of countries' (emphasis added). Unless otherwise stated, all emphasis in quotes has appeared in the original.

15 We shall elaborate more on this issue later (sections 1.3 and 3.1).

16 Abramovitz (1986: 223) argues that '[i]n explaining postwar growth in Europe and Japan, for example, one may be able to say with some confidence that these countries were competent to absorb and exploit then existing best-practice technology' although 'judgements about social capability remain highly problematic'.

17 Freeman (1987: 100); Henderson (1967: 96–118); Landes (1965: 108).

18 Gerschenkron (1962; 1963; 1970); Trebilcock (1981: 205–91).

19 Shin (1992: section 3.4).

20 World Bank (1990; 1991).

21 For instance, the position of dependency theory is summarised as follows:

> Basically the dependence situation is maintained because, in addition to the already stated factors of direct control by the multinationals and dependence on the external markets, the industrial sector develops in an incomplete form. The production goods sector . . . which is the centre pin of accumulation in a central economy, does not develop fully. Ordinarily, economists refer to 'technological dependency' and it means that the economy has to import machines and industrial inputs, and consequently has to stimulate exports (especially of primary goods) to generate the necessary foreign exchange.
>
> (Cardoso, quoted in Fransman 1985: 606)

22 For a general survey of the controversy, see Dahlman and Westphal (1982); White (1978); Pack (1982).

23 Fransman (1985: 606–10).

24 Refer to Shin (1992: section 2.1.2) for a critical account of this.

25 For instance, Abramovitz (1986) deals with this issue when he discusses the case of forging ahead. Ames and Rosenberg's (1963) pioneering work on technological catching-up also deals with various cases of advantages and

disadvantages of latecomers, though their emphasis is different from Abramovitz in that they are more concerned with attacking unfounded contentions of latecomers' advantages.

26 Ernst and O'Connor (1989) also doubt Perez and Soete's contention that it would be easier for developing countries to acquire knowledge and expertise from public domain in the pre-paradigmatic period.

27 One reason for this lack of detailed historical analysis can be attributed to the fact that their main concern is about the impact of new technology, especially that of the ME revolution to the prospects of *current* developing countries. So the debate is still speculative in its nature. The impact of the ME is just beginning to shape, and its precise nature and direction is still moot. This is a backbone of the post-Fordism debate. For a succinct summary of the debate, see Nielsen (1991).

28 A similar account is made by Boyer (1988: 67).

29 Perez and Soete (1988: 459).

30 Perez and Soete (1988: 475).

31 Perez and Soete acknowledge, at the end of their discussion, that:

> any developing country that can truly take advantage of this sort of opportunity [opened by techno-economic paradigm change] has probably reached that position through decades of efforts at entering mature technologies and probably with some successes.
>
> (1988: 477)

But the analysis of how latecomers have reached this position to take advantage of the second window, is missing. It is actually an 'accident' in their framework.

32 This deterministic view of Perez and Soete is accentuated by their confusion between 'catching-up' and 'overtaking'. They identify catching-up as a process of latecomers' gaining net advantages over forerunners in a technology system. Thus, they argue that '[d]evelopment is not about individual product successes but about the capacity to establish interrelated technology systems in evolution, which generate synergies for self-sustained growth process' (1988: 476). This is, in fact, a process of 'overtaking' or 'leapfrogging'. Even if latecomers do not gain technological superiority over their forerunners in a technology system, there are some benefits they can enjoy from exploiting forerunners' technologies, by which they can narrow the gap with forerunners. This would be catching-up in the ordinary sense. It seems to us that Perez and Soete consider only the historical experience of leapfrogging by Germany or the United States in the late nineteenth century owing to their implicit identification of catching-up with leapfrogging. Other cases of narrowing the technology gap within a techno-economic paradigm are neglected here.

33 Refer to Gerschenkron (1962: esp. chapter 2; 1963); Rostow (1971: esp. appendix B); Rosovsky (1965).

34 Gerschenkron (1962: 41).

35 Gerschenkron (1962: 31).

36 Gerschenkron (1963: 167).

37 Gerschenkron (1962: 41).

38 Gerschenkron (1968: 138, emphasis added). In another place, he summarises his position as follows:

> actions by banks and governments in less advanced countries are regarded as successful attempts to create *in the course* of industrialization conditions which had not been created in the 'preindustrial' periods precisely because of the economic backwardness of the areas concerned.
>
> (1962: 358)

Also, he argues that 'the English "prerequisite" when translated into Russian, tended to become . . . a "postrequisite"', when he comments on the cause and effect of 'enclosures' in England and Russia (1963: 168).

39 Gerschenkron (1962: 7).

40 Gerschenkron (1962: 47).

## 2 TECHNOLOGY TRANSFER, TECHNOLOGICAL CAPABILITY AND NATIONAL SYSTEMS OF INNOVATION

1 Rosenberg (1982: vii).

2 For a more detailed discussion on the approach of the neo-Schumpeterians, see Fransman (1985) and Shin (1992: section 2.1.3).

3 Westphal, Kim and Dahlman (1985: 169).

4 Westphal, Kim and Dahlman (1985: 171).

5 Refer to Reddy and Zhao (1990); Lall (1987); Enos (1991); Dahlman and Nelson (1993); Bell and Pavitt (1993).

6 Rosenberg (1982: 271) refers to Rosovsky (1972) and Abramovitz (1986: 222) to Ohkawa and Rosovsky (1973).

7 One difference is the treatment of organisational progress. In Rosovsky's concept of social capability, organisational progress is treated as something independent from technological progress whereas it is dealt with mainly in relation to the latter in the neo-Schumpeterian analysis. This is very similar to the difference between Schumpeter and the neo-Schumpeterians in that the latter focus on the technological aspect of innovation whereas the former's conceptualisation of innovation is broader. Refer to Shin (1992: section 2.1.3).

8 We shall use technological and social capability interchangeably hereafter as far as it does not cause any confusion to our discussion.

9 Perceived in this way, the conceptualisation of technological capability is not really an original contribution of the neo-Schumpeterians. For instance, the same aspect of technological progress is manifested in Kaldor's (1957; 1961) 'technical progress function'. He proposes the technical progress function as a device to explain 'the rate of shift of the production function', i.e., the rate of technological progress. And commenting on the *height* of the technical progress curve, Kaldor (1961: 265) says that it

> expresses society's 'dynamism', meaning by this both inventiveness and readiness to change or to experiment'. In his analysis, 'it is the 'technical dynamism' of the economy, as shown by the height or position of . . . technical progress curve, which is responsible, in a capitalist economy, both for making the rate of accumulation of capital and the rate of growth of production relatively small or relatively large.

However, the major emphasis of Kaldor's technical progress function is placed on the endogeneity of technical progress through its interaction with capital accumulation, not on explicating the height of the technical progress curve. This is mainly because Kaldor, especially in his earlier works, was interested in devising a *general model* of economic growth competing with neo-classical growth models.

10 So, Westphal, Kim and Dahlman (1985) employ the analogy of proficiency when they analyse technological capability. Stiglitz's (1987) concept of 'learning-to-learn' points to the same aspect of technological capability.

11 Refer to Rosenberg (1982); Freeman (1989); Westphal, Kim and Dahlman (1985); Enos (1991); Lall (1987); Bell and Pavitt (1993); Dahlman and Nelson (1993).

12 Some neo-classical economists admit the interaction between technological progress and the increase of inputs. So Solow (1959) and Denison (1964) modify their growth models by incorporating improvements in the quality of the capital stock and labour force, respectively. But these modifications still treat a large part of technological progress as independent from other sources, such as capital and labour. See Abramovitz (1989).

13 Refer to Abramovitz (1989: chapter 1); Nelson (1964); Shin (1992: section 2.1.2). This interaction mechanism was recently rediscovered by new growth theorists such as Romer (1986; 1989); Lucas (1988), and so on.

14 For a fuller discussion of Verdoorn's law, see Kaldor (1966; 1975); Rowthorn (1975a; 1975b).

15 For instance, refer to the quotation from Rosenberg at the beginning of this section.

16 For instance, Westphal, Kim and Dahlman (1985) describe technological progress in a latecomer as a gradual process of acquiring and upgrading technological capability through technological effort.

17 For instance, Enos (1991: 9) argues that '[t]echnological capability may be an input to other economic activities; it is also an output'.

18 For instance, Dahlman and Nelson (1993: 4) argue that investments in social absorption capability 'include, prominently, investment in education across a relatively broad front and, for a certain fraction of the population, to relatively high levels of sophistication in the applied sciences and engineering'. And Abramovitz (1986: 222–3) takes 'years of education' as a 'rough proxy' for technical competence.

19 It should be noted that Rosovsky, who was a student of Gerschenkron at Harvard University, does not treat the initial level of Japanese social capability as a general precondition for economic development. In drawing the 'lessons' from Japan for the current developing countries, he divides Japanese modern economic history into two periods: pre-1900 and the post-1900. He points out the importance of a gap between backwardness in terms of economic indicators and backwardness in terms of other social indicators, i.e., the state of being technologically backward but not so socially backward as its level of technology might suggest, in the pre-1900 period. However, he explicitly states that 'the "initial conditions" in Japan were *peculiar* – compared either to the early industrialisers of the West or to the more backward countries of Asia, Africa, and South America', and argues that 'the pre-1900 period is relatively devoid of "lessons" for the less developed countries' (Rosovsky 1972: 229, 232; emphasis added). Then, he criticises the possible fallacy of attempting to draw general lessons from this period as follows:

> if Japan's early and quite rapid success can be related to a 'gap' between the economic and non-economic state of affairs, then an implication of current interest would follow. Perhaps the emphasis in less-developed countries should be on the creation of this gap through a long-term programme of institution building. To put it another way, on the basis of the Japanese experience, one can preach patience and urge the creation of Meiji-type preconditions. But I find this to be impractical and harmful advice. In effect, this says to countries and policy-makers with urgent problems that a long period of time is necessary to prepare a society for modern economic growth. I prefer to take the position that this suggestion is irrelevant; if this is indeed the lesson of the past, then we must search for substitute patterns in the present.
> (Rosovsky 1972: 236)

20 Ohkawa and Rosovsky (1973: 214). See also Arrow (1962).
21 In this respect, it would be worthwhile to note that capital accumulation is still reaffirmed as a major source of economic growth in advanced countries by many authors (Kaldor 1961; Romer 1989; Scott 1989; De Long and Summers 1991). And the strategic importance of capital accumulation in developing countries has not yet been refuted (Sen 1983).
22 We shall elaborate on this issue later (section 3.1).
23 For fuller discussion of the relation between uncertainty, learning and institutions, refer to Shin (1992: section 2.1.3).
24 For example, the patent system draws the line of appropriability of technologies to some extent. Economic and technological policies of government also affect the perception of uncertainty on the part of private firms. In the same context, Dosi and Orsenigo (1988: 19) argue: '*Uncertainty necessarily implies institutions*, in two senses. First, one requires *behaviour-shaping* institutions . . . Second, uncertainty – even in the weaker form of *imperfect competition* – requires institutions to organise the interactions and the coordination between agents.'
25 For instance, Lundvall (1992: 1) argues, 'learning is predominantly interactive and, therefore a socially embedded process which cannot be understood without taking into consideration its institutional and cultural context'.
26 Nelson (1993: 4).
27 Nelson (1993: 4).
28 Nelson (1993: 19).
29 Lundvall (1992: 18).
30 Lundvall (1992: 18).
31 Refer to section 3.1.1.
32 Lundvall (1992: 1).
33 Nelson (1993: 515); emphasis added.
34 We shall later elaborate on this point in detail (section 3.1.2).

## 3 INSTITUTIONS, INSTITUTIONAL CHANGE AND CATCHING-UP

1 Quoted in Eucken (1950: 59–60).
2 Quoted in Eucken (1950: 55).
3 Thus Menger criticised Schmoller's contention that 'political economy could be reformed . . . by connecting it with historical knowledge', arguing that '[a]ny science can be reorganized only from the inside . . .'. (quoted in Lane and Riemersma 1954: 435).
4 For instance, Fischer (1968: 61) summarises the result of the *Methodenstreit* as follows:

> There can be little doubt that logically Menger was in a far better position. What he advocated as the appropriate methodology for economics was the use of partial-equilibrium models that have since become the common method of theorists in most part of the world. . . . [On the other hand] He [Schmoller] wanted to grasp all at once what has since become the object of many different branches of the social sciences. As a consequence, he touched on many important things, but often only superficially and seldom with clear insight about the implications of his observations and judgements. He was attacked not only by theoretical economists but also by professional historians.

Gay (1954: 411), an economic historian who was a student of Schmoller, also admits that '[t]heir [the Historical School's] criticism has helped to modify the "absolutism of theory", but their work has not displaced that of the "theoretical" school'.

5 Schumpeter (1954: 813). He goes on to say:

> by that time he [Schmoller] had silently unlearned the lessons of extreme 'historicism'. Into a framework that did not depart fundamentally from oldest tradition, he fitted the rich materials of social history. . . . He theorized weakly . . . but he displayed no reluctance to do so. In matters of value and price Schmoller in fact adopted, or meant to adopt, the teaching of Carl Menger.

For criticisms of the method of the Historical School, refer to Eucken (1950: 59–63); Popper (1960); Schumpeter (1954: 809–15).

6 In this respect, Schumpeter (1954: 886–7) argues that

> the general economics . . . was but little affected by these [sociological] developments. Its institutional framework was left practically untouched, that is to say, it was left in the shape in which it had been thrown by the English 'classics' and in particular by J.S. Mill. Nations remained amorphous agglomerations of individuals. Social classes were not living and fighting entities but were mere labels affixed to economic functions (or functional categories). Nor were the individuals themselves living and fighting beings: they continued to be mere clotheslines on which to hang propositions of economic logic. And, with improving rigor of presentation, these clotheslines stood out even more visibly than they had in the works of the preceding periods.

7 Both Schmoller and Menger regarded themselves as economists (Fischer 1968; Schumpeter 1954).

8 Weber (1949: 72).

9 Weber (1949: 75).

10 Quoted in Collins (1986: 35).

11 Weber (1949: 78).

12 Weber (1949: 78).

13 Weber (1949: 72); emphasis added.

14 This tension between the reality and human perception has been a recurrent issue in philosophy and social theories. Kant's dictum that 'concepts without perceptions are empty and perceptions without concepts are blind' (quoted in McCloskey 1991: 96), or the tension between nominalism and realism in social theory, (refer to Smelser 1976: 49) is all concerned with the issue. Eucken (1950: 41) also puts it as 'the Great Antinomy' in that we have to see social phenomena 'as part of a particular individual-historical situation', on the one hand, but we are obliged to see them 'as presenting general-theoretical problems', on the other.

15 For discussions of Weber's ideal type, refer to Weber (1949); Giddens (1971: 133–44); Smelser (1976: 114–50); Eucken (1950: 69–70, 106–9); Schumpeter (1954: 818–19); Commons (1934: 717–48); Gerschenkron (1968: 41–3).

16 Weber (1949: 93).

17 Giddens (1971: 141).

18 Weber (1949: 65). In this connection, he argues that '[i]t is not the "actual" interconnections of "things" but the *conceptual* interconnections of *problems* which define the scope of the various sciences' (1949: 68).

19 Weber, quoted in Giddens (1971: 142).

20 Weber (1949: 79).

21 Weber (1949: 80).

22 Weber (1949: 80); Giddens (1971: 145).

23 Thus Weber (1949) emphasises 'the frequently repeated attempts to discover ever *new* aspects' (97), 'perpetual reconstruction' of concepts (105), or 'transitoriness of all such syntheses' (105). See also Commons (1934: 722, 734).

24 Weber, selected in Runciman (1978: 23).

25 Smelser (1976: 141). According to Smelser, this is what John Stuart Mill called, 'the method of concomitant variation'.

26 Giddens (1971: 139).

27 Smelser (1976: 142). One example is the emergence of the factory system of manufacture in England and Germany. Weber notes some functional 'pre-requisites' for the development of this kind of production system, which include mass market and 'the presence of a sufficient supply of free labourers'. But he generates 'sub-models' which explain the difference between England and Germany in fulfilling the same functional requirements.

> The labor supply in England, for example, was made possible by the eviction of peasants from the land during the enclosure movement, and by the relatively small requirements for manpower for the British armed forces. In Germany shop industries were deliberately established by the state as institutions for relieving and providing work for the poor . . . Demand expanded in Germany because of increased military needs, a new interest in luxuries, and a mass interest in imitations of luxuries.
>
> (Smelser 1976: 126)

28 This method of investigating different causes for the same effect was, according to Smelser, termed by Mill as 'indirect method of difference' or 'joint method of agreement and difference'. It is appropriate to situations when 'the agency by which we can produce [a] phenomenon is not that of one single antecedent, but *a combination of antecedents*, which we have no power of separating from each other, and exhibiting apart' (quoted in Smelser 1976: 142).

29 Smelser (1976: 130) points out that '[m]any of Weber's explanatory "models" . . . were constructed by specifying different combinations and compromises between the "autonomous logic" of ideal types and . . . external determinants'.

30 Smelser (1976: 136); emphasis added.

31 Williamson (1985: 1) boldly argues that economic institutions 'have the main purpose and effect of economizing on transaction cost'.

32 For discussion of different strands of institutionalists, refer to Langlois (1986: 1–25); Hodgson (1991); Matthews (1986); Boulding (1957); Dorfman *et al.* (1963).

33 Williamson (1985: 199); emphasis added.

34 Willamson (1985: 387); emphasis added.

35 Williamson (1985: 18–19).

36 North (1990: 27).

37 Thus Khan (1995) points out that 'the gap between the neoclassical production function and reality can always and tautologically be attributed to transaction cost'.

38 *Ex ante* type transaction costs include 'costs of drafting, negotiating, and safeguarding an agreement', and *ex post* type costs include

> (1) the maladaptation costs incurred when transactions drift out of alignment . . ., (2) the haggling costs incurred if bilateral efforts are made to correct *ex post* misalignment, (3) the setup and running costs associated with the governance structures . . . to which disputes are referred, and (4) the bonding costs of effecting secure commitments.
>
> (Williamson 1985: 20–2)

39 North (1990: 28).
40 Wallis and North (1986); North (1990: 28)
41 For instance, North (1990: 83) asserts that '[t]he agent of change is the individual entrepreneur responding to the incentives embodied in the institutional framework'.
42 So Williamson's (1985) analysis of vertical integration or multidivisional corporations proceeds with the assumption that technology is equal across firms.
43 For instance, see the following remark: 'It is important to stress that . . . either technological change or changes in institutions (political or economic) that affect the specification and enforcement of property rights may increase productivity' (North 1989: 1323).
44 See Chapter 2.
45 North (1990: 84) narrowly argues that these new opportunities, i.e., sources of institutional change, come mainly from changes in relative prices and those of tastes.
46 North (1990: 86).
47 Thus North (1990: 89) contends that 'the single most important point about institutional change . . . is that institutional change is overwhelmingly incremental'.
48 North (1990: 109). He also argues that '[t]he efficiency of the political market is the key to this issue [of persistence of inefficient economic institutions]' (1990: 52).
49 North (1990: 109); emphasis added.
50 He argues that 'on balance nineteenth-century US economic history is a story of economic growth because the underlying institutional framework persistently reinforced incentives for organizations to engage in productive activity however admixed with some adverse consequences' (1990: 9).
51 North (1990: 67). It should be noted that he does not provide the actual comparison of 'transaction cost per exchange' between advanced and Third World countries in his study.
52 For discussion of the second-best theorem, see Bohm (1987); Boadway and Bruce (1984: 131–6).
53 Refer to Rowthorn and Wells (1987); Chang (1994: 56–8).
54 Level effects refer to once-and-for-all changes that raise or lower balanced growth paths without affecting their slope, whereas growth effects refer to changes in parameters that alter growth rates along balanced paths.
55 See Kuisel (1981).
56 Refer to section 3.2.3. and Conclusion.
57 Refer to section 3.2.2 and Conclusion.
58 See Elbaum and Lazonick (1986).
59 Gerschenkron (1962: 50); emphasis added.
60 Gerschenkron (1968: 40).
61 Gerschenkron (1968: 42). In the same vein, he argues in another place that 'concepts are plans for actions: they are programs for research' (1968: 38).
62 Gerschenkron (1968: 42).
63 Gerschenkron (1968: 43).
64 Gerschenkron (1968: 45).
65 Gerschenkron (1968: 46).
66 Gerschenkron (1962: 353–4).
67 Gerschenkron also points out the role played by ideologies in binding the industrialisation process, which is not dealt with in this dissertation. For a discussion of ideology, see Gerschenkron (1962: 22–6; 1968: 57–73).

68 It should be noted that, when Gerschenkron refers to 'institutional factors' or 'institutional instruments', the meaning of institution is not neutral. It connotes a more hierarchical allocation of resources, whereas, in an ordinary definition of institution, both market and hierarchy are included. For instance, North (1990: 3) defines institutions as 'rules of the game in a society', or 'the humanly devised constraints that shape human interaction'.
69 Gerschenkron (1970: 113).
70 Gerschenkron (1962: 9).
71 Gerschenkron (1962: 10, 128–30).
72 Gerschenkron (1962: 360, fn. 11).
73 Gerschenkron (1962: 360). A similar point is made by Pollard (1981: 175).
74 Thus Gerschenkron (1962: 25) argues that 'Buckle was not far wrong when in a famous passage of his *History* he presented the conversion of public opinion in England to free trade as achieved by the force of *incontrovertible logic*' (emphasis added).
75 Gerschenkron (1962: 131) describes the background as follows:

> There is little doubt that military considerations had a good deal to do with the Russian government's conversion to a policy of rapid industrialisation. True, no immediate military discomfiture preceded the initiation of the new policy. But the war of 1877 against the Turks was won on the battlefields in the Danube Valley and the Balkan Mountains, only to be lost in Berlin against the British and probably the Germans as well. In the course of the Berlin congress, particularly during its dramatic moments, the Russian government had much opportunity to reflect that it was not much better prepared for any military conflict with a Western power than it had been a quarter of a century earlier on the eve of the Crimean War. In the short run, Russian reaction consisted in shifting the direction of its expansionist policy away from Europe to Central Asia and the Far East. Taking a somewhat longer view and further prompted by the formation of military alliances in Central Europe, the government turned toward the goal of a drastic increase in the economic potential of the country.

76 Gerschenkron (1970: 102).
77 Gerschenkron (1962, chapter 4).
78 Tilly (1967: 177–80; 1991; 181–2); Landes (1969: 350); Henderson (1975. 113); Trebilcock (1981: 92–104).
79 For a summary of this, refer to Sylla (1991).
80 Abe (1991: viii).
81 Tilly (1967: 179).
82 Gerschenkron (1962: 13–14).
83 Gerschenkron (1962: 14). This is why the German investment banks are called 'mixed banks' or 'universal banks'.
84 Tilly (1967: 178).
85 It should be noted that the formation of business groups is not peculiar to Japan or South Korea. Studies by Hirschman (1968; 1986), Strachan (1976), Leff (1978), Okichi and Yasuoka (1984), and Amsden (1989) view it as rather a general tendency in developing countries, although their size, integrity and efficiency within a national economy vary across countries.
86 Leff (1978: 670).
87 For instance, see the case of Hyundai Heavy Industries (HHI) in South Korea (Amsden 1989: 269–90; Shin 1992, section 3.3).
88 Kim and Lee (1987: 271).

89 For instance, see Feldenkirchen (1981: 32).
90 Gerschenkron (1962: 19).
91 Gerschenkron (1962: 19).
92 Gerschenkron (1962: 12).
93 Gerschenkron (1962: 19). In this passage and another (1962: 8), Gerschenkron expresses these negative roles of the state as 'prerequisites'. But he corrects his position later as follows:

> It took the present writer several years before he succeeded in formulating the concept of prerequisites so that it could be fit into the general approach premised upon the notion of relative backwardness. . . . [therefore] several times the statement is made . . . that abolition of serfdom in Russia was a necessary or even an absolute prerequisite for industrialization. What was meant was that the emancipation of the peasants in Russia of the 1860s very materially aided the subsequent process of industrialization. This is by no means wrong, but it should have been possible to express that thought without mobilizing high-sounding but meaningless terms. There should be a fine on the use of words such as 'necessary' or 'necessity' in historical writings.
>
> (Gerschenkron 1962: 357)

94 Gerschenkron (1962: 19).
95 Gerschenkron (1962: 361).
96 Gerschenkron (1962: 19).
97 Gerschenkron (1962: 19–20).
98 Gerschenkron (1970: 123).
99 Trebilcock (1981: 335–43).
100 For South Asian economy, refer to Khan (1989). For Latin American countries, refer to Hirschman (1968).
101 Gerschenkron (1962: 16).
102 A typical example of this attempt is Weber's (1961) *General Economic History* in which several ideal types Weber had developed were intertwined in explaining the evolution of capitalism. Refer to Collins (1986); Smelser (1976: 114–50).
103 In this respect, Gerschenkron (1970: 126–30) suggests in his later work to view the failure cases of catching-up as a 'counterfactual history'. The Bulgarian experience is one example. In view of Gerschenkron's schema, the positive role of the state should have been expected because of its extreme backwardness during the late nineteenth century. If the industrial take off did not occur in Bulgaria, the schema implies that some factors had prevented its state from performing the positive role. Although the explanation of this reason requires another organising principle, we can at least have a direction for future research, for instance, through following the question why the Bulgarian state actually frittered away its scarce resources instead of devoting them to industrial financing.

## 4 COMPARATIVE SETTINGS

1 Thus the consumption of iron per capita has been regarded as one of the most accurate measures of industrialisation (Landes 1969: 250; Fremdling 1991b: 113).
2 As mentioned before (section 1.1), the post-war economic growth of Western Europe and Japan has been often portrayed as catching-up with the USA, bringing about the Golden Age of Capitalism (see also Abramovitz 1979; 1986; Maddison 1982; 1991).
3 See section 1.1.

4 This sort of view was also expounded by Amsden (1991).
5 Gale (1990: 149–53).
6 Before the invention of the blast furnace, iron was reduced *directly* from iron ore in a bloomery. But with the introduction of the blast furnace, iron is reduced *indirectly*. At first, pig-iron (or cast iron) is smelted (or cast) from the blast furnace. But pig-iron is brittle and not suitable to ordinary usages in itself. So it is converted (or refined) in a furnace called a finery. The puddling process is a converting method which uses hard coal as fuel. Refer to Figure 4.1. for iron- and steel-making processes. It would be worthwhile to note that effective adoption of these new methods was a very slow process in Britain all through the eighteenth century. Darby's invention of coke-smelting was as early as 1709. But it needed 'decades of empiricism' to achieve 'a knowledge of mix and finished product that made it possible to make use of less favourable materials and improvements in the blast had yielded the higher temperature required' (Landes 1969: 89–90). The most serious limitation to the spread of coke-smelting process was relative underdevelopment of the conversion process which was still based on charcoal. This bottleneck was removed only in 1783 by the invention of Henry Cort's puddling process (Gale 1990: 157; Landes 1969: 91).
7 The charcoal-based iron production was effectively dead in Britain by about 1800. One main reason for this transition was due to the earlier depletion of charcoal in Britain (Gale 1990: 156).
8 This was for the following reasons. First, it opened the possibility that the iron production site was not necessarily located in woodlands where a constant supply of charcoal was guaranteed. Second, iron ore and the new fuel, coal, were usually found in the same place, so iron production could be concentrated in coal-producing areas. This locational concentration reduced the transportation costs of raw materials and brought about the development of big industrial districts in South Wales and the West Midlands in Britain and later the Ruhr area in Germany. Third, the application of mineral fuel enabled a drastic improvement of the furnace. The higher temperature attained by coke made it possible to vastly increase the size of the blast furnace. The maximum temperature that can be attained with coke is around 1,000°C, whereas it is about 300°C with charcoal (Gale 1990: 156–9; Landes 1969: 94–6).
9 Landes (1969: 95). This contrasts starkly with the British position in the previous century when it was not an efficient producer in the international market. '[M]uch of the industry survived in the face of Swedish and Russian competition only thanks to customs duties; and production, though rising, was rising far more slowly than imports of the output of more prosperous industries'. Even until the 1780s when the British Industrial Revolution was well under way, Britain's output of iron was smaller than that of France (Landes 1969: 94–5).
10 It should be noted that it is more appropriate to use the term 'the German states' rather than 'Germany' because the German unification was not achieved during this period. But, for convenience, we shall use them interchangeably as far as our discussion is not greatly affected by the choice of the term.
11 Landes (1969: 255).
12 Landes (1969: 259). The mass production of steel made the demand for wrought iron virtually extinct from the late nineteenth century.
13 Paskoff (1991: 78, table 1); Landes (1969: 269).
14 The USA overtaking was primarily based on the introduction of Bessemer-type processes by making bigger and bigger investment as their German counterparts did. One difference from the German case was that the USA exploited both the

original Bessemer process and the Thomas basic process since it had ore mines with both high and low phosphorous contents (Paskoff 1991: 78).

15 See section 3.2.2.

16 In comparison with the steel-making processes before the introduction of the Bessemer process, the open-hearth process brought about a major improvement in fuel economy by 'the utilization of regenerative principle (Landes 1969: 256–7). But, in comparison with the Bessemer process which used no fuel, it still required fuel. This made the process more expensive than the Bessemer process. Moreover, it was much slower than the Bessemer process. Yet, the open-hearth technology had a definite advantage over the Bessemer process since it made steel of much higher quality as 'it enabled the furnace operator to make frequent checks on the steel and to adjust composition as required' (Gale 1990: 172). This quality advantage drove the steel producers to adopt and improve the open-hearth technology.

17 Lynn (1982; 1985); Yonekura (1991: 195–8, 223).

18 Yonekura (1991: 198, 225–6).

19 Yonekura (1991: 215); Hogan (1983).

## 5 REGIONAL CONTRAST IN THE DEVELOPMENT OF THE IRON AND STEEL INDUSTRY IN GERMANY, 1800–1850

1 Tilly (1991: 179); Trebilcock (1981: 92–104).

2 Landes (1969: 180).

3 Landes (1969: 217, table 6).

4 For a broad comparison of the leading industrial regions of continental Europe in the earlier period of industrialisation, see Pollard (1981: 87–106).

5 Landes (1969: 179). According to Landes (1969: 140), the first successful coke-smelting blast furnace was built by William Wilkinson, a British ironmaster, at Ignace de Wendel in France, but the subsequent technological improvement almost stopped in this iron works. Henderson (1967: 17, fn. 2) records that coke-smelting was tried at Sulzbach (Saar) in 1765–1767 but the quality of the iron was unsatisfactory.

6 Pounds (1958: 100).

7 A. Ure, quoted in Henderson (1967: 18).

8 Quoted in Pounds (1959: 195). Apart from iron works, other industries like coal-mining and linen were developed in Silesia. For instance, the first steam engine in German states was introduced in the Silesian coal-mines.

9 Pounds (1958: 100).

10 Henderson (1967: 4).

11 Henderson (1967: 5) describes the involvement of the Prussian state in this region as follows:

> The state reserved the right to designate certain areas as Crown mining concessions. Private persons were allowed to work minerals only after obtaining a concession from the Crown. Ownership of a mine was to be represented by 128 shares (called *Kuxe*). The management of each mine was entitled to the profits of 122 of these shares. Two shares belonged to the ground landlord, two to the local church and school, two to the miners' gild and the poor. At the same time the day to day working of private mines was placed under a considerable measure of State control which was exercised through the officials of the Royal Mining Office.

12 According to Henderson (1967: 6),

> These [skilled worker] immigrants enjoyed various privileges. They were not subject to servile obligations: they paid taxes at a reduced rate; they enjoyed welfare services such as sick pay and subsidised cottages; and they neither performed military service nor provided billets for troops.

13 Henderson (1967: 19).
14 Henderson (1967: 20); Pounds (1959: 195).
15 For instance, Fremdling (1991b: 115) argues that 'they could not compete, even in the area, with iron works that employed traditional methods'. This was mainly because charcoal was still extremely cheap.
16 Pounds (1958: 105); Tipton (1976: 18). In this respect, Pollard (1981: 103) comments that 'all this precociousness was . . . skin-deep only'.
17 He argues that, during the first half of the nineteenth century, the levels of technologies and skills in these areas were so low that autonomous development relying on local technologies was inconceivable. It was necessary to introduce advanced technologies and skills, and this could be done only by the state.
18 The major iron works in this region were spread over the Ruhr, Sauerland and Siegerland. For a detailed explanation of the region, see Henderson (1967: 44–5).
19 For instance, Pounds (1952: 33) describes as follows:

> From the end of the Middle Ages, the Siegerland and Lahn districts had been developing as the primary iron-making and smelting district, while the Sauerland, with its abundant water power, was becoming a centre of the iron-finishing industries.

20 Landes (1969: 217, table 6).
21 Landes (1969: 179) observes that,

> given the differences in quality, charcoal blast iron could compete with the coke-smelting pig; for many, indeed, wood seems to have been almost a free good (save for cost of labour) – if it was not used in the furnaces, it simply went to waste.

22 For a broader discussion of the persistence of old technologies, refer to Rosenberg (1976: 202–6).
23 Landes (1969: 175). Fremdling (1991a: 51) corroborates this view by arguing that '[b]efore 1843/44 the traditional German iron industry had not been significantly or permanently damaged by the new production process'.
24 For instance, the Siegerland, one of the most prosperous areas of the iron and steel industry in Rhineland–Westphalia, was described as follows even in 1846:

> The wooded valleys of Siegerland contained an endless series of little metal working establishments . . . . 'all worked by water power and on the most diminutive scale'. The mines from which the ore came rarely employed so many as ten men. 'Manufacturing . . . is dependent on the land. The furnace owner and forest owner, as well as the miner, club their property together to make iron, living the while on the produce of their little estates'. Iron making in fact was a peasant's by-industry. . . . Having made the iron, they sold it on credit to the small 'hammer master' who beat and refined it into steel and sold it to the working cutler; who sold to the dealer; who sold to the shipping house. Perhaps, at these last two stages, something which might be called capitalism cropped out.
>
> (Clapham 1921: 89)

25 Fischer (1963: 89); Henderson (1975: 212).
26 Henderson (1984: 41); Fremdling (1977: 595; 1991b: 117).
27 Also see Fremdling (1991a: 62, table 2.2); Landes (1969: 217).
28 Crouzet (1972: 123).
29 According to Pounds and Parker (1957: 116), the Krupp works in Essen and the workshop of Jacob Mayor in Bochum 'were achieving an international reputation for their prices of cast steel' during the period.
30 Landes (1969: 178).
31 Studies assessing the degree of protection given to the German wrought iron sector by the Maassen tariff are not available to my knowledge. But there are some clues for the assessment. According to Fremdling's (1991a: 61, table 2.1) calculation, the ratio of British wrought iron exports to pig-iron exports to Germany and Holland area was 43.4 per cent in 1821 and the ratio rapidly reduced to 4.8 per cent in 1830, 1.4 per cent in 1840, and 0.7 per cent in 1850. And, in the 1820s, the puddling process began widely spreading in Rhineland–Westphalia. This suggests that the German wrought iron sector was on the road of progress in spite of the import of British wrought iron.
32 Crouzet (1972: 123) also observes that, owing to the differential tariff system, 'the making of iron by puddling developed a great deal in the 1830s and 1840s, using imported British or Belgian pig-iron'.
33 Fremdling (1991b: 117).
34 See section 4.2.
35 Quoted in Crouzet (1972: 121). These concepts are similar to Hirschman's (1958; 1968) 'backward linkage' and 'forward linkage'.
36 Fremdling (1991a: 50–1).
37 This was in fact a kind of 'capital stretching' frequently found in many successful late industrialising countries in their attempts at reducing capital–labour ratio. See Rosenberg (1982: 271–2).
38 Fremdling (1977: 595). Here, we deliberately emphasised the advantages of the German iron and steel industry as a late-comer. This does not mean that the German iron and steel industry enjoyed only advantages compared to its forerunners. One important disadvantage seems to have been the amount and prices of fuel. Although woods were abundant in Germany during the nineteenth century, they were not so abundant and cheap as coals in Britain. So Landes (1969: 181) stresses the importance of the incentive for technological progress in fuel economy as follows:

> The fact remains, however, that continental iron-masters were making more of their resources than their competitors across the Channel; and since fuel economy was the key to efficiency in almost every stage of manufacture, these tentative advances of the 1830's and 1840's were the starting-point of a scientific metallurgy that was to pay off in major improvements a generation later.

39 Freeman (1989); Lee (1991: 3).
40 Henderson (1967: 21–57).
41 Diefendorf (1980: 322, 326).
42 Böhme (1978: 1) describes the situation of the German territories at the end of the eighteenth century as follows:

> about 1800 customs barriers hindered trading relations in central Europe. In Prussia alone there were 67 local customs tariffs and just as many customs borders. . . . [there were] the 314 sovereign imperial territories and the 1,475 imperial knights whose world was governed by the infrastructure provided by the Holy Roman Empire of the German nation.

43 Henderson (1967: 21).
44 For instance, Diefendorf (1980: 213) notes that 'Rhinelanders had been faced with an indeterminate period of military requisitions, forced loans and emergency taxes, and the quartering of troops'.
45 For instance, Vicke's proposal for major road-building in Westphalia, which later significantly contributed to the development of heavy industries in the Ruhr, was delayed by lack of state funds. 'It was only after a second large loan had been successfully floated in London (1822) that the Prussian government was able to allocate substantial funds for the building of highways' (Henderson 1967: 48).
46 Diefendorf (1980: 318–22, 316); Pollard (1973: 642).
47 For instance, the French form of local government on the left bank of the Rhine survived even until 1845. And many businessmen who worked for the local government under the French rule, were allowed to retain their offices (Diefendorf 1980: 213–88).
48 In this respect, the successful role of the Prussian state had to do with the increasing governability of this newly acquired region. For instance, Henderson (1967: 47) describes Vicke's achievement in Westphalia as follows:

> By 1844 the horizon of the inhabitants of Westphalia had widened. The petty provincialism of the eighteenth century – with its restricted loyalties to small territories such as the Mark or Münster – had given way to a wide patriotism. The citizens of Essen, Dortmund, Münster and Minden were now all Westphalians and Prussians.

## 6 UK–GERMANY–JAPAN COMPARISON IN THE IRON AND STEEL INDUSTRY, 1850–1930

1 Landes (1969: 218) describes the rapid extinction of the traditional methods in the 1850s as follows:

> For Prussia as a whole, representing about 90 per cent of the iron production of the *Zollverein*, the proportion of charcoal-blast iron fell from 82 per cent in 1842, to 60 per cent in 1852, to 12.3 per cent in 1862.

2 Fremdling (1991b: 121).
3 For a broad comparison of the two countries, see Black *et al.* (1975, esp. 16, 119 20).
4 Gerschenkron (1962: 10).
5 Henderson (1975: 79).
6 Trebilcock (1981: 77).
7 Fischer (1963: 88) points out that it freed 'mining enterprises from the state supervision' and reduced government functions 'to those of special police and technical supervision'.
8 Landes (1965: 151).
9 Cameron (1956: 119); Tilly (1967: 162).
10 Tilly (1967: 163).
11 Chandler (1990: 416).
12 Fremdling (1977); Trebilcock (1981: 60).
13 Chandler (1990: 490).
14 Feldenkirchen (1981: 31).
15 Feldenkirchen (1981: 31).

16 Fremdling (1991b: 123) also points out that the new British plants erected in the mid-1870s 'boasted of newer equipment' than their German counterparts.
17 According to Landes (1969: 264), by 1890, 'the average German open hearth was half . . . as large as the British', but, in 1901, 'German converters poured out an average of 34,000 tons . . . where British equipment produced 21,750'.
18 Landes (1969: 263–9).
19 Tilly (1967: 169).
20 Tilly (1967: 170–4).
21 Tilly (1967: 169).
22 In general, not just confined to the iron and steel industry, Schumpeter describes the role of the German investment banks as follows:

> [They] . . . provided machinery to do this themselves. They took care of the necessary issues of stocks and bonds, thus helping the enterprise to redeem its short debt and providing it with additional means. In order to effect this they were ready to take these stocks or bonds for their own account, not only if they were unable to place them, but in the ordinary course of their business routine. . . . When eventually they placed the securities acquired, they again financed the private investors so that, temporarily at least, the transaction often meant no more than a shift in asset.
>
> (quoted in Tilly 1967: 178)

23 Feldenkirchen (1981: 29).
24 Feldenkirchen (1981: 31).
25 Thus Henderson (1975: 180) calls cartels 'children of financial storm'. But it should be noted that some steel producers organised a cartel without the bank's influence in the 1860s, albeit on a very small scale and for a short period. During the Depression, cartelisation became an economy-wide phenomenon due to the banks' influence.
26 Fremdling (1991b: 125).
27 According to Fremdling (1991b: 125),

> it was finally the power of the great banks that forced Krupp into the cartel. The Dortmunder Union . . . threatened Krupp with cutthroat competition by the railmakers association. Due to his capacity and the high productivity of his Bessemer plant, Krupp did not have to fear any competition based solely on productive power. As he was not backed by any major bank, he would have lacked the financial power to survive cutthroat competition, so he finally joined the cartel in 1876.

A similar bank enforcement of the cartel took place in 1904 when Phoenix held out from the Steel Mill Federation (Webb 1980: 312; Trebilcock 1981: 97).
28 Landes (1969: 245); Fremdling (1991b: 140); Chandler (1990: 423).
29 For instance, Fremdling (1991b: 140) puts it thus:

> Government toleration or even promotion of syndicalization should be regarded as an instrument of national economic policy. Within the ruling elite, the opinion prevailed that the moderate pricing policies of cartels created stability and helped to maintain high employment – thus protecting the national labor.

30 Webb (1980: 313).
31 According to R. Allen's (1979: 928) calculation,

> [in the early 1880s] American rails sold at 13 percent above cost; British rail at 19–20 percent. German domestic prices exceeded costs by 24 percent,

but export prices equalled only 92 percent of costs. The high price realized on domestic sales was enough to make German rail production as a whole profitable.

32 For instance, Webb (1980) and R. Allen (1979) argue that the German producers were on average 10–15 per cent more efficient than the British producers around the turn of the century. But Fremdling (1991b: 132–3) rejects their arguments because those calculations do not include the product differentials coming from the fact that British producers were specialised in high-quality open-hearth steel whereas Germany producers were specialised in low- and medium-quality Thomas steel.

33 Webb (1980: 326).

34 Schumpeter's (1943: 104) emphasis on the beneficial effect of monopoly on innovation touches the same point:

> Perfect competition implies free entry into every market . . . But perfectly free entry into a *new* field may make it impossible to enter at all. The introduction of new methods of production and new commodities is hardly conceivable with perfect – and perfectly prompt – competition from the start.

35 Fremdling (1991b: 125) observes, though he is sceptical about Webb's and Allen's quantitative figures (quoted in n. 32), that,

> roughly half of the Bessemer plants were out of blast in 1877. But the remaining plants were used intensively. The entrepreneurs and managers were well aware that the average costs of producing steel declined with the enlargement of the operating scale in a certain plant.

36 Webb (1980: 318).

37 For instance, the effect of mechanical loading alone was 'to reduce a work force of 46 skilled and unskilled labourers per hearth to 16 and cut labour costs . . . by 58 percent' (Landes 1969: 265).

38 Webb (1980: 319).

39 They are larger furnaces, power-generation from waste gas, mixers for molten pig iron, soaking pits, continuous rolling mills, and so on (Webb 1980: 323).

40 Quoted in Trebilcock (1981: 70).

41 A similar feature appears in oligopolistic competition in Japanese *zaibatsu* (or *keiretsu*) or Korean *chaebols*, as shall be elaborated later (sections 7.1, 8.3 and 8.4).

42 The portion of the light industry in the total industrial output was 81.3 per cent in 1878, 80.9 per cent in 1895, 62.0 per cent in 1920 and 58.8 per cent in 1930 (Rosovsky 1961: 99, table 27).

43 Rosovsky (1961: 98–9).

44 One important factor would be Japan's loss of tariff autonomy until the 1890s, which forced the country to engage in free trade from the beginning of its industrialisation. On the other hand, Japan was very cautious of introducing foreign capital, as will be discussed later. The combined effect was that it was imperative for Japan to have competitive export industries to earn foreign currencies if it wanted to import capital goods and raw materials for its industrialisation. Light industry, especially the textile industry, was more promising than heavy industry for attaining international competitiveness in a short time (Rosovsky 1961: 101). This contrasts with the situation of Russia, which relied heavily on foreign investment retaining its tariff autonomy. Another factor for the difference of the industrial structure between the two countries is

related with the availability of industrial labour. In Russia, 'labour supplies for industry were scarce, unskilled, and intractable – and, therefore, in real terms *expensive*' (Trebilcock 1981: 236). By contrast, 'Japan was capable of forming a disciplined and reliable labor force, suitable for factory work, more easily than many backward countries of continental Europe' (Rosovsky 1961: 103).

45 Yawata state iron works dominated the Japanese iron and steel industry producing 74 per cent of pig-iron and 81 per cent of steel products in 1914 (Okazaki 1991: 168, table 1).

46 Most modern industrial ventures started as state enterprises though an extensive privatisation process soon followed in the 1880s. Even in this earlier state-initiated industrialisation, however, 'the government continuously sought to induce private capital into the new industrial fields, but . . . efforts to do so were conspicuously unsuccessful' (Smith 1955: 36).

47 According to Morikawa (1982: 138), 'its pig iron contained so much copper that it was not suitable for steel-making'.

48 Morikawa (1982: 138).

49 Morikawa (1982: 142). According to Nagashima (1991: 191–2)

> [b]efore World War I, the Railway Ministry, and the Army and Navy bought about half of the steel products . . . which Yawata . . . produced; for example the rails produced by Yawata, the advantage to which the Railway Ministry gave over imported rails, were the greatest output of steel products of Yawata before World War I. But the rails of Yawata did not meet the standards of the Railway Ministry. Yawata could not produce high-quality rails until the latter half of the 1920s.

50 As Morishima (1982: 97) points out,

> [t]he consistent aim of successive governments from the Meiji Revolution onwards was to build Japan into a strong country with a top class military capability and a top class industry – a country which could not be defeated by the advanced nations of Europe and America.

See also Hirschmeier (1964: 147).

51 Quoted in Smith (1955: 38). This contrasted to the case of the Japanese modern textile industry which developed largely by private sector though its initial beginning was supported by the state (Smith 1955; Rosovsky 1961: 100).

52 Morikawa (1982: 139).

53 Okazaki (1991: 173).

54 This had to do with applying Japanese raw materials to the imported modern equipment which was designed and installed by German engineers (Morikawa 1982: 140).

55 Okazaki (1991: 174) describes it as follows:

> As is evident in the fact that the process of raw material input was not mechanized until 1914, operations depended highly upon manual labor, and labor productivity was low, roughly one-third or half of that in developed countries. Additionally, blast-furnace gas and coke-oven gas were not utilized for steel-making and rolling; therefore coal consumption per ton of finished steel was 3.3 t, compared with a maximum of 2.7 t in Europe.

56 Allen (1981a: 82). Russia had already attained 45 per cent of self-sufficiency ratio in pig-iron in the 1870s, although most of them came from old charcoal furnaces. The output of iron and steel was increased by fourfold in Russia during the Witte

era from 1893–1900. By 1900, 'Russia had ousted France from the fourth rank in world iron production and had taken the fifth rank in steel manufacture' (Trebilcock 1981: 233, 225).

57 Okazaki (1991: 169). But the importance of the tariff system in the development of the Japanese iron and steel industry does not seem to have been so important as the cases of Germany and Russia. As Allen (1981a: 133–4) puts it:

> To sum up the effects of the tariff, it can be said that until 1920 it had little influence on the economic structure of the country, for before 1911 Japan's freedom to impose protective duties was limited by treaty, and between 1914 and 1920 normal trading conditions were so disturbed as to reduce them to a position of minor significance. Even during the next decade, the effects of the tariff on foreign competition in the home market was overshadowed by exchange fluctuations, and it was never more than a secondary weapon in the armoury of economic policy.

58 Smith (1955: 98).
59 Lockwood (1954: 322).
60 Landes (1965: 96).
61 Trebilcock (1981: 235).
62 Landes (1965: 96–7). The contribution of the foreign capital to its industrial development was more than what this figure implied because foreign capital was 'heavily concentrated in certain regions and sectors, invariably those upon which rapid growth was most dependent' (Trebilcock 1981: 245).
63 Trebilcock (1981: 224, emphasis added). For instance, the mills in Ekaterinoslav, established by a foreign entrepreneur, Hughes, produced half of Russia's steel output in the 1890s.
64 Landes (1965: 98–9).
65 Landes (1965: 100); Hirschmeier (1964: 131).
66 Trebilcock (1981: 227–8).
67 Morishima (1982: 93).
68 Smith (1955: 86–100); Morishima (1982: 94).
69 Gerschenkron (1962: 22).
70 One private integrated producer, Toyo Seitetsu, 'entrusted Yawata Seitetsujo with the management of its iron works' in 1921, and another private integrated works, Tanaka Kozan, was bought by Mitsui Zaibatsu in 1924. 'As a result, all enterprises using continuous process of producing pig iron came under state ownership (management) or Mitsui/Mitsubishi Zaibatsu ownership'. Yawata embarked on rationalisation of its own iron works and entrusted Toyo iron works with the funds paid from the government general account in addition to its own income. Mitsui/Mitsubishi supplied the funds from their affiliated firms like Mitsui Bank Co., Mitsui & Co., Mitsubishi Goshi (the holding company of Mitsubishi Zaibatsu), and so on (Okazaki 1991: 182–3).
71 For instance, the technology for producing high-quality basic pig-iron was first introduced at Mitsubishi Seitetsu early in the 1920s and Yawata followed suit in the mid-1920s (Okazaki 1991: 184).

## 7 USA–JAPAN–KOREA COMPARISON IN THE IRON AND STEEL INDUSTRY, 1950–1990

1 Tiffany (1991: 246–7; 251).
2 Paskoff (1991: 82–9). Hard driving was a method of intensifying the operation of

the blast furnace. It increased the production volume of pig-iron by about 50 per cent over the previous level although it shortened the life of the blast furnace lining.

3 Paskoff (1991: 91).

4 U.S. Steel 'began its corporate life with nearly 66 per cent of domestic ingot production, and its share remained at 33 per cent just prior to the start of World War II (when its capacity was two and one-half times more than its nearest rival)' (Tiffany 1991: 248).

5 According to Tiffany (1991: 249), Gary, the chairman of U.S. Steel during the period, 'was more interested in achieving a stable stream of profits to satisfy his shareholders and obtaining peace with government, rather than pursuing cost reduction through innovation and efficiency'.

6 Tiffany (1991: 250–1) describes the situation as follows:

> U.S. Steel, realizing its loose and generally uncoordinated structure was too burdensome in this period, finally began to restructure its manufacturing facilities in 1935. But after more than 30 years of operations, this giant firm could not reorder its 'corporate culture' so easily or quickly.

7 Tiffany (1991: 252).

8 Lynn (1982: 157–65; 1985: 259).

9 Yonekura (1991: 198, table 4).

10 Tiffany (1991: 256).

11 In order to spur rapid expansion of steel production demanded by the war, 'the government provided various financial incentives to the producers', but 'there was a strict time deadline that had to be met to obtain these public benefits, thus pressing the firms to *move quickly*' (Tiffany 1991: 255; emphasis added).

12 According to Yonekura (1991: 197–8), '[i]t reduced refining time to a tenth of what it had been, thereby increasing the efficiency of the furnace, and it saved half the construction cost'. See also Lynn (1982: 16, fn. 2).

13 Compared with the existing OHF technology whose efficiency came largely from the extensive use of scrap, BOF technology was characterised by 'lower or no utilization of scrap' (Lynn 1982: 39–40; Yonekura 1991: 223).

14 Yonekura (1991: 224). According to Lynn (1982: 19), Defasco, a small Canadian steel-maker, solved some technical problems 'to a level of marginal satisfaction through extensive pilot experiments between 1952 and 1954'.

15 Yonekura (1991: 225).

16 Lynn (1982: 37); Yonekura (1991: 198; 225–6).

17 For instance, Japan 'imported 133.7 million tons of ore, which was over 99 percent of its requirements' and '58.3 million tons of coking coal, or 90 percent of its requirements' in 1980 (Hogan 1983: 67).

18 Yonekura (1991: 214).

19 Yonekura (1991: 215); Hogan (1983: 65); Allen (1965: 161).

20 For instance, Hogan (1983: 69) points out that,

> [B]y 1965, a number of these areas [with extensive reserves of high-grade ore in Australia] were being developed with capital made available based on long-term ore contracts with the Japanese; in fact, it could be said that the Australian iron-ore industry got its start from the Japanese commitments to take large tonnages over protracted periods.

21 For instance, it was estimated that the specialised large cargo ships alone would reduce transportation costs by 20 per cent (Yonekura 1991: 223). Also refer to Hogan (1983: 75); Allen (1965: 161).

22 Yonekura (1991: 197, table 3). McCraw and O'Brien (1986: 404, n. 20) describes the increased scale of the Japanese iron and steel industry as follows:

> in 1976, Japan had thirty-seven blast furnaces with an inner volume measuring in excess of 2,000 cubic meters compared to only five in the United States. . . . In 1980, the average output of Japan's blast furnaces was 2.3 million tons a year while the average for US blast furnace was .84 million tons.

23 Japanese producers could undercut domestic prices of the USA from the 1960s by up to 30 per cent 'even after expensive transportation costs were absorbed' (Tiffany 1991: 256). This competitive export price was also partly due to 'export dumping' (Allen 1965: 164, fn. 17). Japan became the largest steel exporting country in the capitalist world in 1973 (McCraw and O'Brien 1986: 87).
24 Abe (1991: ix). If other export items made of steel like automobile and ships are counted, the export ratio would be much higher. For instance, Hogan (1983: 83) argues that '[i]n total, it is conservatively estimated that one-half of Japanese steel production is destined for export, either as steel products or in the form of products containing steel.'
25 Even in general industrial development, Patrick and Rosovsky (1976: 47) argue that 'the main impetus to growth has been private . . . Government intervention generally has tended (and intended) to accelerate trends already put in motion by private market forces'.
26 Refer to Yonekura (1991: 213–14).
27 Yonekura (1991: 219).
28 Thus, Kawasaki's construction of blast furnaces 'was aloofly regarded by the government as *an exceptional case*', probably because it secured the World Bank loan. Other producers were not allowed to build blast furnaces (Yonekura 1991: 219, 217).
29 Yonekura (1991: 221–2). In this context, Yonekura (1991: 204) argues that 'Interpretations such as "MITI targeted several industries for potential export growth: steel was one of them", were only made in hindsight.'
30 The Okano Plan, which provided a basic idea for heavy and chemical industrialisation, was initially rejected by Prime Minister Yoshida in 1953. The plan was adopted only during the period of the five-year plan (1955–1960) which coincided with the Second Rationalisation Programme in the iron and steel industry (Johnson 1982: 228–31).
31 Refer to E.B. Schumpeter (1940); Allen (1940; 1965; 1981a); Morishima (1982).
32 For instance, Miyashita and Russel (1994: 33) point out that '[w]hen a portion of their managers were actually purged, this merely provided opportunities for new executives . . . to step forward'.
33 This was also the point made by Abramovitz at the level of the overall economy (section 1.1).
34 It came into being in 1934 after the merger between Yawata and six other private producers, the main purpose of which was 'to extend government control over industries of strategic importance' in establishing a war economy. The Japanese government held most of the capital in this new company and the various *zaibatsu* the rest (Allen 1965: 155).
35 Yonekura (1991: 213).
36 Yonekura (1991: 213).
37 Yonekura (1991: 219).
38 In the post-war Japan, there emerged two kinds of *keiretsu*: the horizontal *keiretsu* and the vertical *keiretsu*. There are six big horizontal *keiretsu*: Mitsui,

Mitsubishi, Sumitomo, Fuyo, Sanwa, and DKB (Dai-Ichi Kangyo) Groups. Most of them, except Sanwa Group, have their lineages with the pre-war *zaibatsu*. Whereas the *zaibatsu* took the form of 'family concerns' linked vertically and topped by a holding company for the whole group, the post-war horizontal *keiretsu* lacked the family control and holding companies. Instead, the latter is mostly centred around commercial banks. The relation between banks and other companies, and that between companies are more voluntary and take the form of a loose federation. The horizontal *keiretusu* system is often identified as the whole Japanese *keiretsu* system mainly because of their sheer size and their lineage from pre-war *zaibatsu*. But there are also vertical *keiretsu*, which are centred around a big parent company by subcontracting networks. The leading vertical *keiretsu* are Toyota, Nissan, Honda, in the automobile industry, and Sony, Matsushida, Hitachi, Toshiba in the electronics industry. As will be elaborated in this section and later (section 8.2), the horizontal *keiretsu* played a major role in the expansion of the iron and steel industry whereas the vertical *keiretsu* played a more significant role in the semiconductor industry. Refer to Miyazaki (1980a: 55); Miyashita and Russell (1994); Kiyonari and Nakamura (1980: 255).

39 For instance, in 1959, 'the average share of retained earnings in the capital investment of the ten largest American and four largest German firms was 91.7 per cent and 93.1 per cent respectively, while that of Japan's six largest was 45.6 per cent' (Yonekura 1991: 227).

40 The two *keiretsu* were holding stakes of the former Japan Steel. When Nippon Seitetsu was formed in 1934, Mitsui's Wanishi iron works and Kamaishi iron works, and Mitshbishi's iron works were merged (Morikawa 1982: 134).

41 Miyashita and Russell (1994: 99).

42 But it should be remembered that the relation between individual firms and the bank has not been a completely exclusive one. See Kiyonari and Nakamura (1980: 253).

43 Miyazaki (1980a: 1980b); Kiyonari and Nakamura (1980). In pre-war Japan, the feature of oligopolistic competition was not so strong, although there was a strong rivalry among different *zaibatsu*. Each *zaibatsu* had some kind of specialisation (Miyazaki 1980a: 70). In the iron and steel industry, the specialisation across *zaibatsu* was more evident owing to the dominant position of Yawata (Morikawa 1982: 141).

44 For instance, Yonekura (1991: 228–35) emphasises the ineffectiveness of MITI's attempts at cartelisation whereas McCraw and O'Brien (1986) and Johnson (1982) argue for the effectiveness of MITI's intervention.

45 Section 6.1. Also refer to Fremdling (1991b).

46 Nippon steel had market shares of 44.5 per cent of pig-iron and 35.4 per cent of crude Steel in Japan when it started.

47 Even the president of NKK who had to compete with the newly born giant, commented as follows:

> I have been thinking that the best way to control such excessive competition is through the merge of Yawata and Fuji. If they behave well as a leader of the industry in the competition over capacity expansion and sales, we will find a way out of this morass.
>
> (quoted in Yonekura 1991: 235)

48 For instance, Pempel points out that '[m]acro-level economic growth has been the central *political goal* to which all other Japanese policies have been subordinated during the post-war period' (quoted in Borrus, Millstein and Zysman 1983: 184, emphasis added).

49 Cumings (1987: 64).

50 Johnson (1982; 1987). It should be noted, however, that the emergence of the developmental state does not mean that the capability of the state sector became superior to that of the private sector. It only means that the state's pursuit of industrial development became less obstructed by other political goals or growth-retarding political pressures. As discussed previously, the Japanese private sector had been developed enough to initiate catching-up process in the iron and steel industry. Moreover, in comparison with its follower, Korea, the Japanese developmental state's capability was much more constrained than the Korean, as will be discussed later (section 7.2).

51 For instance, the drastic expansion of the iron and steel industry during the 1960s was fuelled by the ambitious 'income doubling plan' launched in 1960. In order to achieve its macro-level goal, the Plan 'forecast that crude steel capacity would increase 48 million tons by 1970, and thus 30 new blast furnaces would be needed'. The Japanese producers responded to this plan more than the state expected. The industry invested ¥1,092 billion ($3 billion) and crude steel capacity increased more than 70 million tons by 1970 (Yonekura 1991: 218, 226).

52 Zysman (1983: 247); Ueno (1980: 403); Johnson (1982: 203–7); Borrus, Millstein and Zysman (1983: 195).

53 Industrial policy can be defined as 'a policy aimed at *particular industries* (and firms as their components) to achieve the outcomes that are *perceived by the state* to be *efficient* for *the economy as a whole*' (Chang 1994: 60). For the importance of industrial policy in Japanese economic development, refer to Johnson (1982); Ueno (1980); Allen (1981a; 1981b); Dore (1986); Freeman (1987); Okuno-Fujiwara (1991).

54 Ueno (1980: 388–9). Zysman (1983: 248) also puts it as follows:

> the market for industrial lending is structured by the government. Prices do not clear the market . . . . the level of interest rates was kept low with the deliberate aims of reducing the cost of manufactured exports and stimulating investment. *Prices in all but the very short-term money markets are artificially maintained by government.*

55 JDB's contribution to total industrial capital was 22 per cent in 1953. But it was soon reduced to 7 per cent in 1957 and 4 per cent in 1960 (Johnson 1982: 212, table 15).

56 Yonekura (1991: 219–20). Also see Johnson (1982: 200); Ueno (1980: 401–2).

57 It should be noted, however, that the Japanese state's role in credit allocation was, although more direct than the German state's intervention, less direct than that of its follower, Korea, reflecting its moderately backward economy as compared to Korea's 'extremely' backward economy (section 7.2).

58 Yonekura (1991: 219–20).

59 Those measures included,

> the limitation of imports by imposing quotas on foreign currency, directly limiting the amount of imports by such measures as the introduction of an import licensing system, and indirectly regulating imports by such means as imposing a high protective tariff on imports and introducing a preferential commodity tax for domestically produced goods.
>
> (Morishima 1982: 188–9)

60 McCraw and O'Brien (1986: 96).

61 According to McCraw and O'Brien (1986: 405, n. 38), 'in April 1951, the domestic delivered price of steel plates in Japan was $137 per ton, compared to $82 per ton

in the United States. . . . Japanese steel became price-competitive with American-made products in about 1958.'

62 Johnson (1982: 278).

63 We can only point out that the combination of the Cold War international politics and the USA's dominant position among the capitalist countries would be one principal reason for it. As Millstein (1983: 112) argues, '[g]iven the strategic exigencies of postwar reconstruction in both Europe and Japan, the United States tolerated the maintenance of high tariff barriers in Europe and the imposition of restrictive import quotas and foreign direct-investment controls in Japan'.

64 Lynn (1982: 51–4; 1985: 258–60); Yonekura (1991: 224). Peck and Tamura argue that 'overall effect of MITI intervention in the mid-1960s was to provide Japanese firms with technology on more favorable terms than they would have been able to obtain in the more open economies' (quoted in Fransman 1985: 629).

65 MITI called for '*keirets*ification' in 1953 no sooner had the SCAP's ruling ended. And it helped regrouping of trading companies through 'its licensing powers and ability to supply preferential financing', and 'ultimately winnowed about 2,800 trading companies that existed after the occupation down to around 20 big ones, each serving a bank keiretsu or a cartel of smaller producers' (Johnson 1982: 206). Also see Miyashita and Russell (1994: 37).

66 Thus, even the World Bank, which generally holds a sceptical stance on the efficacy of industrial policy, acknowledges its effectiveness in Japan especially in 'industries with scale economies' as a special case. Refer to World Bank (1993).

67 POSCO accounted for around 80 per cent of domestic steel production in 1990 (Juhn 1991: 274). And it is currently the sole integrated producer in Korea.

68 In 1962, when Korea began its industrialisation in earnest, the crude steel-making capacity was just 14,000 tons per year which supply only 3.5 per cent of domestic demand. With the rapid industrialisation in the 1960s, several non-integrated mills were established. But most of their production facilities were small and outdated. The self-sufficiency ratio in steel rose only to 23 per cent in 1972 just before POSCO began its production (Juhn 1991: 269; Hogan 1990: 29).

69 Hogan (1990: 29). POSCO rose to the second largest steel producer only after Nippon Steel and Korea became the sixth largest steel producing country in 1993 (*Maeil Kyungje* – The Economic Daily 3/9/1994).

70 POSCO formed a joint venture with USX (United States Steel) for the purpose of modernising USX's Pittsburg, Pennsylvania, plant (Amsden 1989: 291–2).

71 POSCO's initial plan to build a plant with a capacity of 0.6 million tons, which was less than the internationally competitive scale of around 1 million tons, failed mainly because the IBRD assessed even that scale was too big for Korea and suggested that Korea should give priority to the machinery industry rather than to the iron and steel industry. KISA (Korea International Steel Associates), an international consortium formed by Western steel-makers, abandoned the project due to the objection from the IBRD. When the new round of negotiations began with Japan in 1969 after the failure with KISA, POSCO's president, T. J. Park, proposed to the Japanese collaborators that the plant should be producing at least 2 million tons per year, which was the production capacity of the newest furnaces at that time. Although the final agreement for the scale of the initial mill was reduced to 1.03 million tons, POSCO extracted agreement from the Japanese that the capacity of the plant should be expanded to 3 million tons within the next three to five years. Moreover, when POSCO was going to build the fourth plant, T. J. Park persuaded the Japanese that it should have a capacity of 9.1 million tonnes – not the 2.6 million figure that was the basis of the first plan

received from Nippon Steel's technical cooperation department (POSCO 1989: 129–56; Amsden 1989: 295–6; Enos and Park 1988; 177–80).

72 Around 1990, the rate of CC was 95 per cent at Pohang plants and 100 per cent at Kwangyang plants (Juhn 1991: 283).

73 Juhn (1991: 283) points out that '[t]he average age of major facilities at POSCO is only 6.1 years, compared to 17.8 years at major integrated steel mills in Japan'. In particular, Kwangyang complex is currently referred to as 'the most modern and largest in the world' (Hogan 1990: 31).

74 It is commonly held that the Korean industrial take-off took place from the 1960s. Its annual GNP growth rate had been slightly more than 4 per cent in the 1950s and it increased to more than 9 per cent from the beginning of President Park's regime (1961–1979). In terms of aggregate and industrial growth rates, there was a definite discontinuity (Cole and Lyman 1971; Mason *et al.* 1980; Jones and Sakong 1980; Song 1990). Nevertheless, historical continuity should not be ignored. For instance, Cumings (1987: 57) argues that 'Korea's industrial revolution' already began during the 1930s and early 1940s when Japan mobilised Korean resources, both human and physical, to the utmost for its war purpose. Despite the fact that the Korean peninsula was divided into North and South Korea after its liberation from the Japanese occupation in 1945, and that most of the industrial facilities were kept in the North, subsequent South Korean industrialisation did not begin from nothing (see Mason *et al.* 1980: 292). South Korea also underwent a short period of slow import-substituting industrialisation in the 1950s.

75 Amsden (1989: 292).

76 POSCO (1989: 156); Cole and Lyman (1971: 134, table 6.9).

77 POSCO (1989: 156, table 1.5.1).

78 The composition of POSCO's initial funding of $138.7 million of foreign currency requirement was as follows: reparations ($30.8 million), Japanese Export Import Bank loans ($50 million), government loan ($42.9 million), and government investment ($15 million) (POSCO 1989: 156).

79 POSCO (1989: 148).

80 Refer to Johnson (1987).

81 Cumings (1987: 67).

82 This was mainly because President Park, a military man favouring efficiency through a 'top-down' process rather than political consensus, relied on bureaucracy rather than any political party. The government bureaucracy was modernised by the influence of the military which was then the most 'modernised' institution in Korea (Cole and Lyman 1971: 36).

83 Mason *et al.* (1980: 253–4) describes the Rhee regime's (1948–1960) characteristic as follows:

> There was, then, no dearth of plans and planning activity before 1961. The reason why this activity had so little effect on policy was, essentially, that President Rhee was more interested in other things than in economic development. As in so many of the new states, the leader who fought for independence proved not to be a man capable of effective administration. . . . Thus his anti-Japanese measures retarded the resumption of trade with a natural and traditional partner, and his yearning for unification was carried to the extreme that there was an unwillingness to build up the South as an independent and integrated economy.

84 For a detailed account on this, refer to Cole and Lyman (1971: 78–97, esp. 80); Yoon (1990: 71–3). For instance, see the following remark by President Park himself:

> In order to ensure efforts to improve the living conditions of the people in Asia, even undemocratic emergency measures may be necessary . . . It is also an undeniable fact that the people in Asia today fear starvation and poverty more than the oppressive duties thrust upon them by totalitarianism. . . . In other words, the Asian peoples want to obtain economic equality first and build a more equitable political machinery afterwards. . . . The gem without luster called democracy is meaningless to people suffering from starvation and despair.
>
> (quoted in Mason *et al.* 1980: 251)

85 Johnson (1982: 228). See also section 7.1.

86 For instance, Suh Suk Joon, then Deputy Vice-Minister of EPB (Economic Planning Board) recalled that,

> the HCI policy was necessary [to solve the BOP problem] because the ingrained nature of the economy was such that the dependence on foreign supplies of intermediate goods was too high, the value-added productivity was too low, and the upgrading of export industries was in an unsatisfactory state.
>
> (quoted in Yoo 1990: 25)

87 From 1961 to 1968, when export-oriented industrialisation based on light industrial goods was well under way, total imports almost doubled total exports (Cole and Lyman 1971: 128, table 6.4). This BOP problem continued until the mid-1980s, resulting in heavy accumulation of foreign debts (Amsden 1989: 94–6).

88 US protectionism in the early 1970s was mostly directed towards labour-intensive products, especially towards textile which had been the single most important export item of Korea. The bilateral textile agreement with the USA in 1973 was a 'heavy shock' to the Korean business community and the government. Moreover, the so-called second-tier NICs in the South East Asian countries like Thailand, Indonesia and Malaysia were beginning their catching-up process in the light industries (Yoo 1990: 20–1).

89 North Korea dispatched a group of specially trained soldiers to Seoul in an attempt to take President Park's life in January 1968, and a few days later they seized the U.S.S. Pueblo, a US navy vessel and its eighty-three-man crew in the open sea (Yoo 1990: 19).

90 Nolan (1986: 62).

91 President Park explicitly expressed this kind of nationalist vision in one of his speeches:

> Economic resurgence is an integral part of nationalist vision of a more independent Korea to come – more independent of the United States aid and control and, as an economically stronger and independent entity, more able to deal with North Korea.
>
> (Park 1963, quoted in Mason *et al.* 1980: 46)

92 But it seems that economic imperatives were stronger than military ones. For instance, Oh Won Chul, one of the most important architects of the HCI programme, recalled that 'defense needs and economic considerations were given weights of 20 per cent and 80 per cent, respectively' (quoted in Yoo 1990: 19, n. 8).

93 See Jones and Sakong (1980); Chang (1990); Shin (1992, chapter 3); Yoo (1990).

94 Trebilcock (1981: 84).

95 The establishment of POSCO was included in the Second Five Year Plan (1967–1971) while the other major HCI programme was included in the Third Five-Year Plan (1972–1976).

96 Chang (1990: 24–5).

97 The import liberalisation ratio in 1975 was 67.1 per cent for iron and steel, while it was 35.0 per cent for chemicals, 33.3 per cent for fabricated metal, 33.8 per cent for non-electric machinery, and 17.6 per cent for electric machinery (Yoo 1990: 39, table 7). Heavy protection on the machinery sector was maintained even up until the late 1980s (Jacobsson 1993).

98 Refer to Byun (1980: 128, table 3).

99 Amsden (1989: 301). For instance, in the case of hot-rolled steel product, '26.5 per cent . . . was exported, 39.5 per cent was sold to local consumers that were eligible for duty drawbacks, and only 34 per cent was sold to consumers that would have had to pay a 25 per cent duty had they imported steel' in 1983 (Paine Webber, quoted in Amsden 1989: 301).

100 In fact, it sometimes set export prices higher than domestic prices when the international market was booming. For instance, its average export price of steel products in 1974 was $299 per ton while its domestic supply price was $256 per ton. In 1978, its export price exceeded domestic price again. The former was $284 per ton while the latter was $261 per ton (Byun 1980: 128, table 3).

101 For a detailed account of the pricing policy of POSCO, refer to SCRC (1987: 236–56).

102 The overall policy direction of the Korean state can be termed as 'EP (export promotion)-*cum*-IS (import-substitution)' (Shin 1992, section 3.1; Amsden 1991). We shall elaborate on this strategy in the case of the semiconductor industry later (sections 8.2 and 8.3).

103 POSCO raised foreign long-term loans at a low rate from international banks without any payment guarantee from the government. One example was 'a 6.95 per cent loan to be repaid over 20 years from the Japanese Export–Import Bank'. And 'Domestic finance of Kwangyang's first phase, estimated to be $1.37 billion of a total investment cost of $1.96 billion, was covered entirely by POSCO retained earnings' (Juhn 1991: 277).

104 POSCO (1989: 334). This was a 'shock' to the Japanese steel producers who had been the main supplier of technologies and capital for POSCO. Subsequently, they refused to give assistance to establishing POSCO's new steel complex at Kwangyang (POSCO 1989: 333–6).

105 Most of the infrastructure for POSCO was provided by the government. According to one estimate, the government spent around $42 million, which amounted to over one-fifth of the initial investment of POSCO, just for the 'massive supporting facilities'. POSCO was allowed to generate its own electricity to save electricity bills outside the government electricity regulations. The state 'also provided POSCO with discounted user rates for many government services, such as a discounted railroad rate of 40 per cent, port rate of 50 per cent, water-supply rate of 30 per cent, and gas rate of 20 per cent' (Amsden 1989: 297). The significance of government subsidies on the overall competitiveness of POSCO is, however, still contentious. For instance, Juhn (1991: 275) argues as follows: 'Although some critics contend that heavy government subsidies were the major factors of POSCO's low cost structure, most experts conclude that even after adding in subsidies to POSCO's costs, the company was still competitive with Japanese steel-makers'.

106 The capacity utilisation ratio was almost 100 per cent from the beginning of its operation in 1973 until the middle of the 1980s, while it was 55 per cent for American producers and 62 per cent for Japanese producers during the same period (SCRC 1987: 271–2; Enos and Park 1988: 215).

107 According to Amsden (1989: 296–7),

> POSCO managed to complete construction of the first phase of its mill to a mere twenty-seven months . . . Of a total of twenty-seven facilities, twenty-three were completed *ahead of schedule* . . . In fact, delays in construction of a hot-strip mill led to a crash project in which, with round-the-clock construction, workers completed five months work in two months . . . A comparison study . . . indicates that construction costs per metric ton of integrated mills after the oil crisis were the following: $1,750 for Brazil, $820 for the United States, $700 for the European Economic Community, $590 for Japan, and roughly $400 for POSCO.

See also Enos and Park (1988: 188, table 7.5) and Juhn (1991: 278, table 3).

108 Hogan (1990: 32); Enos and Park (1988: 215–16); POSCO (1989: 335–6).

109 Juhn (1991: 282) points out that, as a result of several intensive efforts to reduce the time for the absorption of imported technologies, 'the time taken from start-up to normal operations averaged only three months for Pohang, compared to 6 to 12 months for the steel mills in developed countries'.

110 We suspect that this globalisation process during the post-war period was one reason why trade was and is 'an engine of growth' whereas it had been only 'a handmaiden of growth' in the nineteenth century (refer to Kravis 1970; Riedel 1984). In a similar context, Kennedy (1989: 537) refers to the emergence of 'trading states' during the post-war period.

## 8 USA–JAPAN–KOREA COMPARISON IN THE SEMICONDUCTOR INDUSTRY, 1950–1993

1 In the neo-Schumpeterian long-wave theorists' terms, steel was a 'key factor' in the third Kondratiev cycle (1880s and 1890s to 1930s and 1940s) while semiconductors are a key factor in the newly anticipated fifth Kondratiev cycle (1980s and 1990s to ?) (Perez 1983; 1985; Freeman, Clark and Soete 1982; Freeman and Perez 1988; Freeman 1992).

2 For instance, Morgan and Sayer comment that 'semiconductors are now accorded a strategic role for national economies similar to that of steel fifty years ago' (quoted in Henderson 1989: 4).

3 Kim, Hong and Lee (1992: 122).

4 A major breakthrough in this system-like character was the invention of the microprocessor in 1971 by Intel. As 'a computer on a chip', the microprocessor provided 'an opportunity to employ a single device as a building block in a vast number of different applications by devising different instruction sets' (Uenohara *et al.* 1984: 10) (also refer to Rosenberg 1982: 184; Freeman, Clark and Soete 1982: 119).

5 For instance, entire production systems were rendered obsolescent after only 2–3 years in dynamic random access memory (DRAM) production (Gregory 1986: 210). And one commentator puts it, exaggerating rather: 'You almost have to invest a dollar to sell a dollar's worth of integrated circuits' (quoted in Gregory 1986: 94).

6 A similar point was made by Suzuki (1991).

7 Soete (1985: 417); Freeman (1992: 87).

8 But recent development of the electronic mini-mills might have a potential to challenge the validity of this argument in the future. See Gale (1990: 184–5).

9 Henderson (1989: 32).

10 This would be partly attributed to the fact that there existed a fierce competition in the US semiconductor industry whereas this competitive pressure was almost absent in the US iron and steel industry during the twentieth century.

11 In the same vein, Dosi (1984) notes that 'diffusion might not often reach 100 per cent: given a very fast rate of technical progress, other innovations are likely to supersede the previous ones before they reach a complete diffusion' (quoted in Soete 1985: 421).
12 Perez (1985: 448).
13 In the same vein, Freeman (1987: 77) argues that 'the competitive pressures in the world economy are so strong that it is very hard to be "non-conformist" once a paradigm has crystallised'.
14 Refer to Conclusion.
15 Henderson (1989: 49–76). So original equipment manufacturing (OEM) and own design and manufacture (ODM) are becoming rules, rather than exceptions in the electronics industry.
16 It was Japan, in particular the introduction of the transistor radio by Sony in 1955, that made the transistor and products using it popular, although the transistor was originated from the USA (Uenohara *et al.* 1984: 14). But, in terms of R&D expenditure and number of patents, the USA was definitely the technological leader (Dosi 1984).
17 Hobday (1989); Cho (1995).
18 For instance, in the 1950s, 'government R&D contracts accounted for approximately 60 per cent of total expenditures at IBM, and government purchases of data-processing equipment for military and space programs provided a ready market for successive generations of computers' (Gregory 1986: 6).
19 Uenohara *et al.* (1984: 14).
20 Refer to Fransman (1990: 13–38).
21 Japanese catching-up effort which was coordinated between the government and private sector in the early 1960s was stalled with the introduction of IBM System 360 in 1964. The catching-up effort during the latter half of the 1960s was also not successful. 'It was only in 1968 that Fujitsu completed the FACOM 230-60 which it introduced in order to cope with IBM's 360 series which reached the market in 1964.' But IBM once again forged ahead with the introduction of its System 370 in 1970. Even major US computer manufacturers like RCA and General Electric began abandoning their computer divisions, unable to compete with IBM any longer (Fransman 1990: 23–38).
22 Gregory (1986: 6).
23 Uenohara *et al.* (1984: 15).
24 Borrus, Millstein and Zysman (1983: 206).
25 Sobel, quoted in Fransman (1990: 49).
26 Borrus, Millstein and Zysman (1983: 207).
27 Fransman (1990: 57).
28 Borrus, Millstein and Zysman (1983: 208).
29 Borrus, Millstein and Zysman (1983: 209); emphasis added.
30 Borrus, Millstein and Zysman (1983: 207); emphasis added.
31 Weinstein *et al.* (1984: 39).
32 Weinstein *et al.* (1984: 39); Fransman (1990: 83).
33 Uenohara *et al.* (1984: 17); Fransman (1990: 81–96).
34 Hobday (1989: 235–6).
35 Gregory (1986: 214–15).
36 Flaherty and Itami (1984: 169–70); Dosi (1984); Methé (1992).
37 Weinstein *et al.* (1984: 50).
38 According to Flaherty and Itami (1984: 165), 'the Japanese companies went from spending about one-third the US companies' capital expenditures in 1973 to about one-half in 1980'.

39 Gregory (1986: 95, 209). The same pattern of differing investment behaviour between US and Japanese producers was repeated in the 1985–1986 recession: whilst US producers shrank from investment, the Japanese kept their pace of investment (Hobday 1989: 235–6).

40 Weinstein *et al.* (1984: 39, 52–3, 70).

41 Okimoto (1984: 78).

42 Fransman (1990: 13–23).

43 This was reflected in government subsidy for R&D,

> Prior to 1972, the percentage of government R&D support was less than 2 percent of the total; in 1973 it increased to about 15 percent . . . Government support peaked at about 26 percent in 1977, reflecting both the increase in funds for the VLSI project and the stagnation of industrial investment resulting from a slowdown of the economy. . . . Government support of R&D dropped to less than 2 percent after the end of the VLSI project in March 1980.
>
> (Uenohara *et al.* 1984: 22)

According to one estimate, between 31 per cent and 51 per cent of total information technology R&D was undertaken under the auspices of MITI-initiated cooperative research programmes between 1972 and 1976 (Flamm, quoted in Fransman 1990: 49).

44 On the prewar development of the Japanese telecommunication and electric power industry, refer to Fransman (1990: 13–14).

45 According to Okimoto (1984: 109),

> Japan's tariff rates for ICs stood for many years at 15 percent . . . In 1972, Japan's effective rate dropped to 12 percent; by 1980, as Japanese industry became more competitive, the rate was brought into line with America's at 10.1 percent.

46 Borrus, Millstein and Zysman (1983: 175) describe the Japanese quota system as follows:

> prior to 1975, integrated circuits that contained more than 200 circuit elements could not enter Japan, from any point of origin, without the permission of the Japanese government. . . . [after the elimination of quota system in 1976 with 'liberalisation'] Japanese customs officials began . . . to treat imports from the US affiliates in less developed countries as coming directly from the United States if more than 50 per cent of an item's value-added originated in the United States.

47 IBM Japan, a wholly owned subsidiary of IBM, was an exception because it was established in 1950 during the Occupation period. But 'IBM computers manufactured in Japan were treated as foreign computers' (Fransman 1990: 303, n. 36).

48 NTT procurement policy, which was completely closed to foreign firms and did not allow Japanese producers to use imported semiconductors in the telecommunication equipment to be supplied, was a positive factor for the development of the Japanese semiconductor industry. And the establishment of Japan Electronic Computer Corporation (JECC) was an important demand-side boost for the Japanese computer manufacturers facing a stern competition from IBM (Borrus, Millstein and Zysman 1983: 215; Okimoto 1984: 109–10).

49 Borrus, Millstein and Zysman (1983: 175).

50 Borrus, Millstein and Zysman (1983: 176). Texas Instruments (TI) story is most illuminating in this regard:

> Texas Instruments petitioned the Japanese government for a wholly owned subsidiary in the early 1960s, and was offered a minority-share joint venture, which it rejected. Its chief bargaining chip during these negotiations was its continuing refusal to license its critical IC patents to Japanese firms without gaining in return a substantial production subsidiary in Japan. NEC and the other firms sub-licensed to it were in fact producing ICs based on technology developed by TI and Fairchild, through an NEC–Fairchild licensing agreement. However, because the TI–Fairchild patent accord explicitly excluded Japan, those Japanese firms were not protected, as Fairchild licensees in Europe, against patent-infringement suits brought by TI. The Japanese government stalled approval of TI's patent application in Japan, this enabled NEC and other firms to play domestic technology catch-up, thereby forcing TI to negotiate for quicker access. The Japanese government then held up Japanese exports of IC-based systems to the United States because TI threatened infringement action. A compromise was finally reached in which TI got a 50 percent share of a joint venture with Sony. In return, it was agreed to license its IC patents to NEC, Hitachi, Mitsubishi, Toshiba, and Sony, and agreed further to limit its future share of the Japanese semiconductor market to not more than 10 percent.
>
> (Borrus, Millstein and Zysman 1983: 203–4)

51 The US government certainly supported its semiconductor industry to a very significant extent as we mentioned above. But it was less organised since most of the support came from the military sector 'without direct reference to the development of the industry' (Borrus, Millstein and Zysman 1983: 150). Putting it more simply, 'Washington had no counterpart to Japan's . . . MITI' (Okimoto 1984: 87).

52 As was the case in the iron and steel industry, the ratio of external financing was high in the semiconductor industry. For instance, external financing as a percentage of total finance was 45 per cent in the Japanese semiconductor industry between 1975 and 1979 whereas it was 21 per cent in the USA industry (Flaherty and Itami 1984: 138).

53 Borrus, Millstein and Zysman (1983: 199).

54 Borrus, Millstein and Zysman (1983: 197–8), Flaherty and Itami (1984: 138–40), and Gregory (1986: 132–7) argue that the Japanese overborrowing system also provided the Japanese industry with lower-cost capital. Their contention is mainly based on supply-side factors. That is, they argue that the separation of Japanese financial system from the international market until the early 1980s and the Japanese constant higher savings ratio, contributed to lowering the cost of capital in Japan. If their contention is correct, this advantage in the cost of capital provided Japanese producers with a great advantage in a prolonged race of investment in the semiconductor industry. But they do not answer the demand-side questions as to whether or not higher demand for investment in Japan increased the cost of capital in the domestic market as compared to that in the international market. In particular, they do not answer the question why Japanese firms were eager to raise foreign capital in the 1970s and in the early 1980s when the Japanese Ministry of Finance still exercised strict control on capital inflows.

55 Fransman (1990: 52). For a detailed account of risk-hedging through the division of specialisation in Japan, refer to Fransman (1990: 19–56).

56 For example, a manager of NEC admitted that 'his firm would have had to spend five times as much on the development of electron-beam technology without the VLSI program' (Borrus, Millstein and Zysman 1983: 211).

57 Borrus, Millstein and Zysman (1983: 211).
58 Gregory (1986: 26, 436).
59 Millstein (1983: 117–18).
60 Freeman (1987: 38).
61 This classification came from Miyashita and Russell (1994).
62 NEC was the most prominent case in point (Miyashita and Russell 1994: 51).
63 For an account of their broad business concerns, see Borrus, Millstein and Zysman (1983: 189).
64 Whilst the share of semiconductor sales to total sales of major US semiconductor producers ranged from 31 per cent to 89 per cent in 1979, that of Japanese producers ranged from 2.3 per cent to 17.8 per cent in the same year (Borrus, Millstein and Zysman 1983: 188).
65 The fact that major Japanese steel-makers have attempted to diversify their businesses into the semiconductor industry, and POSCO is diversifying into telecommunications rather than mechanical engineering or other steel-consuming industries, testifies to the loose technological linkage between the iron and steel industry, and its upstream or downstream industries (Elsevier Advanced Technology 1991; *Maekyung Business Week*, 9 Feb. 1994).
66 But it should be noted that, although IBM and AT&T were artificially blocked in entering the semiconductor business, it is often pointed out that the initial development of the US semiconductor industry was supported by close technological and demand-side interaction with IBM and AT&T (see Dosi 1984; Borrus, Millstein and Zysman 1983).
67 In the same vein, Okimoto (1984: 89) points out that '[w]ith the US concentrating on military and industrial applications, and Japan lagging behind in state-of-the-art technology, Japanese companies had little choice but to find niches in the consumer electronics market'.
68 For instance, in 1981, 'the consumer electronics industry took about 50 percent ($2.2 billion) of Japan's total production of semiconductors ($4.4 billion), a figure far in excess of the US, where consumer goods accounted for only around 15–20 percent' (Okimoto 1984: 89).
69 For the case of NEC, see Gregory (1986: 69).
70 Borrus, Millstein and Zysman (1983: 191).
71 Uenohara *et al.* (1984: 25); Gregory (1986: 45).
72 Uenohara *et al.* (1984: 25). Flaherty and Itami (1984: 154) also observe that '[w]hen they got their bank loans, they included the semiconductor finance needs with those of the rest of the corporation'.
73 Thus Flaherty and Itami (1984: 175) argue that 'for Japanese semiconductor firms as a group [and as a member of *keiretsu*] there was a greater possibility of maintaining financing during a prolonged "race" than for US firms, all else equal'.
74 This dual structure was common in the electronics industry of other Asian NICs like Taiwan, Singapore and Hong Kong, although the degree of protection of their domestic market varied across countries. Refer to Henderson (1989: 49–76).
75 Kim (1980: 4); Mody (1990: 297).
76 Kim (1980: 5).
77 Kim (1980: 11).
78 Kim (1980: 12).
79 Suh (1975: 112, table 6).
80 Other sectors in the electronics industry were developed by local firms or through joint ventures.
81 Yoon (1990: 70); Kim, Hong and Lee (1992: 245).
82 For instance, only 6 per cent of MNCs in Korea were wholly owned subsidiaries

whereas that share was 50 per cent in Mexico and 60 per cent in Brazil (Evans 1987: 208).

83 Yoon (1990: 77).

84 Yoon (1990: 76).

85 For instance, domestic production of radios was promoted by the state in the early 1960s through an import ban on radios and the state's campaign to send radios to rural areas. Samsung and Daehan's entry into the electronics industry in the late 1960s resulted from the state's promotion of large-scale vertically integrated manufacturers (Yoon 1990: 75–6, 79).

86 Its founder was a politician who did not have any prior experience in the electronics industry. It was reported that his beginning of the semiconductor business was due to the advice from the then minister of MCI and Anam received special benefits from the government. (Yoon 1990: 84–5).

87 Yoon (1990: 93). In this context, the Korean electronics producers responded to a survey that '[g]overnment and multinational firms were important sources of external influence on technological development' until the early 1970s (Kim 1980: 9).

88 Consumer electronics which had undergone an import-substituting process during the 1960s, consisted of only 29 per cent of production and 22 per cent of export in the electronics industry by 1973 (Suh 1975: 111, table 5).

89 Daehan was later acquired by Daewoo group.

90 Another reason for the retreat of foreign direct investment was related with the growing difficulty in using developing countries as 'export platforms' due to the increased protectionism in the advanced countries (Kim 1990: 161–2).

91 Amsden (1989: 265); Yoon (1990: 93).

92 Yoon (1990: 98). The first chip produced by Samsung in 1975 was a watch chip. And it developed transistors for monochrome TV and audio in 1976, and ICs for colour TV in 1981. Refer to Choi (1991: 35–6); Archambault (1991: 32–3); Yoon (1990: 105).

93 Yoon (1990: 105).

94 For a broad description of the Korean state's promotional measures for the semiconductor industry in the mid-1970s, see Yoon (1990: 93–5).

95 Kim and Lee (1987: 284).

96 Yong-Chan Park (1987: 105).

97 Kim and Lee (1987: 284).

98 Yong-Chan Park (1987: 107); Archambault (1991: 32–3).

99 Yong-Chan Park (1987: 101).

100 Yoon (1990: 93).

101 *Far Eastern Economic Review*, quoted in Archambault (1991: 35; emphasis added).

102 For instance, even the Law for Importation of Foreign Capital amended in 1988 'clearly states that technological licensing is banned in industries where local capability is deemed to be promising' (Chang 1994: 113).

103 In this respect, Yoon (1990: 101) argues that 'the development process of the Korean semiconductor industry in the latter half of the 1970s illustrates a limitation of an internally strong state constrained by external factors'. His remark is pertinent even in the 1960s, as we pointed out before.

104 Samsung developed 64K DRAM in 1983 with a design bought from a small US firm. Samsung's commercial production of 64K DRAM began in 1984. Hyundai and Goldstar began producing 64K DRAM from 1985 (Archambault 1991: 58; Choi 1991: 40–2).

105 Moreover, Samsung is beginning to lead some segments of semiconductor

technology. According to Elsevier Advanced Technology (1993: 151), 'At the end of 1992, Samsung claimed it had taken the lead in DRAMs by producing a 64 Mbit DRAM chip in which all the storage cells worked first time.' And Samsung announced in 1994 that it developed 256M DRAM first in the world (*Maeil Kyungje – The Economic Daily* 30 Aug. 1994).

106 Choi (1991: 16); Dataquest (quoted in *Maeil Kyungje – The Economic Daily* 19 Feb. 1994).

107 Apart from DRAM production, Korea's catching-up in assembly would be worth mentioning. Anam has remained the largest assembler in the world from the early 1980s. And Hyundai is currently the second-largest assembler in the world. But the level of technological sophistication is much lower in assembly than in wafer fabrication. Anam started the assembly from 1970 based initially on low wages of Korea, and then countered the wage increases by introducing automation and foreign direct investment. Anam is also expanding its business to wafer fabrication. The case of Anam would be a typical import-substituting process, which deserves further research on its own merit. But, the discussion below will be focused on wafer fabrication. Although Hyundai has increased its assembly capacity dramatically, its main effort at catching-up in the semiconductor industry was placed on wafer fabrication. And the portion of assembly in Korea's overall semiconductor production has been decreased substantially from around 80 per cent in 1986 to around 30 per cent in 1994 (Chu 1992: 28, table 3.2; Cho 1995).

108 The Korean government announced in 1981 that its promotion measures concerning the electronics industry would be concentrated on the semiconductor sector. It also worked out 'The Long-Term Semiconductor Industry Promotion Plan (1982–1986)' (Yoon 1990: 106).

109 As mentioned above, the state's regulation on the introduction of foreign technology and capital was significantly relaxed. Tax reductions were extended even to foreign firms as a way to induce advanced technologies. The government embarked on several R&D activities in the broad field of the semiconductor industry. Tariffs were exempted in the imports of semiconductor equipment from 1983 and a 20 per cent flat-rate import-duty was placed on semiconductors (Y.B. Kim 1991: 164; Choi 1991: 46–7; Archambault 1991: 33).

110 It was triggered by the increasing difficulty in securing advanced technologies. US and Japanese producers became very reluctant to sell their technologies as the Korean producers had been successful in catching-up in 256K DRAM and 1M DRAM with imported technologies.

111 Choi (1991: 48; 52–4); *Maekyung Business Week* (17 Aug. 1994).

112 The share of the top ten *chaebols*' value-added to GDP was more than doubled from 5.1 per cent in 1973 to 10.9 per cent in 1978 (Jones 1987: 101, table 3). The average number of affiliated firms of the top ten *chaebol* increased from 7.5 in 1972 to 25.4 in 1979 (Cho 1991: 184–5). And there was a positive correlation between the growth rate of *chaebol* and their degree of concentration on the heavy and chemical industries (Jones 1987; Shin 1992, section 3.3).

113 Choi (1991: 61); Kim, Hong and Lee (1992: 238). Thus, a director of Samsung remarked:

> If you take it only in business terms, I don't think we would get into this business. But the problem is that without semiconductor support, electronics, telecommunications and low-level computer businesses will reach a plateau where they can't make further progress.
>
> (cited in Archambault 1991: 57)

114 Yoon (1990: 109).
115 For instance, Samsung made its semiconductor firm (then Samsung Semiconductor and Telecommunications (SST)) public in 1984 to raise funds for its investment (Yoon 1990: 112).
116 Yoon (1990: 112).
117 Yoon (1990: 111).
118 Yoon (1990: 109–10).
119 Yong-Chan Park (1987: 109). Thus he argues that 'however important the research associations [organised by the state] are in bringing about structural dynamism in industrial R and D, they represent only a supplement to the large corporations' R and D' (1987: 63).
120 Samsung's export of 64K DRAM began only in September 1984. But international prices of 64K DRAM dropped sharply from the end of 1984, mainly due to Japanese firms' dumping. The price of 64K RAM was $3.5 in the middle of 1984. But it dropped to $1.0 by the end of the year and to $0.30 in August 1985. It was slightly recovered to $0.60 by the end of 1985. Samsung's production cost for 64K DRAM by the end of 1985 was $1.40, which means it had to produce DRAMs with accumulating losses (Choi 1991: 44).
121 Yoon (1990: 113).
122 Yoon (1990: 112); Choi (1991: 44); Archambault (1991: 61).
123 Archambault (1991: 78); Choi (1991: 45–6); Yoon (1990: 113).
124 Choi (1991: 44–5); Yoon (1990: 112).
125 Yoon (1990: 113); Choi (1991: 46).
126 The research was undertaken in a competitive base, that is, individual participants were to receive funds differentially according to their respective contribution to the research results. So one could argue that individual firms participated in the collaborative research in order to gain financial benefits provided by the state. Thus, Choi (1991: 53) concludes that the state's organisation of the collaborative researches at least contributed to facilitating individual firms' R&D activities.
127 Borrus, Millstein and Zysman (1983: 210).
128 Kim, Hong and Lee (1992: 242). The pace of import-substitution in the equipment thereafter also has been very slow. Korea still depended on imports for 88 per cent of equipment in 1994 (*Maeil Kyungje – The Economic Daily* 22 May 1994).
129 Refer to our discussion on late vs. late-late industrialisation in the Conclusion.
130 It increased from 19 per cent in 1977 to 22 per cent in 1980 and 32 per cent in 1988 (Choi 1991: 15).
131 For instance, 63 per cent of wafer fabrication products went to the export market in 1986 (Choi 1994, table 4.2).
132 In the same vein, the Korean computer industry started basically as an export industry in the early 1980s. 'Approximately 60% of micro- and personal computers and over 80% of peripherals produced in Korea in 1985 were exported, mostly for OEM buyers under buyer's brandnames' (Kim and Lee 1987: 279).
133 Kim, Hong and Lee (1992: 246–72); Choi (1991: 15).
134 Archambault (1991: 60).
135 According to Kim and Lee (1987: 288),

> [t]hese outposts are manned primarily by Korean–American scientists who had worked at American computer firms such as IBM, DEC, Bell Labs and Hewlet-Packard and who now design computers and semiconductor memory chips. The California offices also serve as an 'antenna' for

information on research activities and training posts for scientists and engineers from R&D centres and manufacturing plants in Korea.

136 Archambault (1991: 36).
137 Choi (1991: 39).
138 Chu (1992: 36, figure 3.3).
139 Okimoto (1984: 91, table 6). This trend still remains. In 1993, the ratio of DRAM production was 20.4 per cent in Japan, 6.6 per cent in the US, and 71.9 per cent in Korea, whereas it was 16.2 per cent in the world market (Cho 1995).
140 *Far Eastern Economic Review*, quoted in Archambault (1991: 58).
141 Archambault (1991: 78).
142 Quoted in Archambault (1991: 79).
143 Of the fifty largest *chaebol* in 1990, the owner did not coincide with the chairman in only one group, KIA. (Cho 1991: 38–9, tables 1–4).
144 For instance, the total sales of Samsung Group was $21.1 billion in 1987 (Mody 1990: 295, table 3). And those of Hitachi and Toshiba were $23. 6 billion and $20.2 billion respectively (KEIPA 1988: 173, 175; $ = ¥ 123.5 at the end of 1987).
145 Thus, Lee Kun Hee, the chairman of Samsung, remarked in an interview:

> It's difficult to catch-up with the Japanese in every area, so we are concentrating our R&D investment on semiconductors – a key industry needed for the development of both Samsung and Korea – and the aim is to bring our level of technology in this area up to the most advanced levels found anywhere.
>
> (cited in Archambault 1991: 64)

Samsung's R&D structure reflects the chairman's commitment. 'In 1990, R&D spending invested in by the semiconductor division was almost Won 170 bil (US$ 240 million), which accounted for 18 percent of its sales, and approximately 40 percent of the total R&D spending of SEC [Samsung Electronics Co.]' while the share of R&D investment to total sale of SEC in the same year was 8.5 per cent (Koh 1992: 29).
146 The ratios of investment to sale for five Korean wafer manufacturers were 225 per cent in 1985, 73 per cent in 1986, 52 per cent in 1987 and 1988, 66 per cent in 1989, and 25 per cent in 1990 (Choi 1991: 82, table 82).
147 Choi (1991: 20).

## CONCLUSIONS

1 The page reference is to Hirschman (1968).
2 Emphasis added.
3 Hirschman (1968: 92–3).
4 Hirschman (1968: 91–2).
5 Hirschman (1968: 90).
6 Hirschman (1968: 112).
7 Hong Kong and Singapore, as city countries, did not have any base for raw materials and agriculture. Korea and Taiwan underwent a short ISI period in the 1950s. But these processes were sustained on the US aid of raw materials and foods. Once the generous US aid was discontinued, they were forced to export (Hong 1993: 187–8; Mason *et al.* 1980; Song 1990; Deyo 1987).
8 Rhee, Ross-Larson and Pursell (1984); Gregory (1986: 411–40).
9 Dahlman and Frischtak (1993: 428).
10 Dahlman and Frischtak (1993: 432).

11 L. Kim (1991b); Shin (1992); Hong (1993); Song (1990).

12 It should be noted that our argument is different from the neo-classical contention of efficient resource allocation through the 'getting prices right' effect of export markets. In fact, East Asian countries, especially Korea and Taiwan, employed lots of deliberate policies of price distortions (Amsden 1989; Wade 1990; Chang 1994; Shin 1992).

13 Wilmot argues in this respect that 'successive waves of Japanese . . . firms targeted US and European markets – each with the aim of achieving a given percentage of world sale' (cited in Hobday 1989: 228).

14 According to Hobday (1992: 82), '[i]n 1980, for example, the share of external component sales in total revenue was less than 5 per cent for Phillips, Siemens, Thompson, AEG–Telefunken and Plessy'.

15 Labour productivity of European manufacturers was estimated 'between less than one third and half of the Japanese level' in 1978 (Dosi 1981: 185). And Europe's share of world semiconductor sales fell from 16 per cent to 10 per cent between 1980 and 1988 (Hobday 1992: 82).

16 Hobday (1992).

17 Mody (1990: 293).

18 Hou and Gee (1993: 397–9).

19 For theories of balanced growth, see Rosenstein-Rodan (1943); Nurkse (1953); Scitovsky (1954). And for those of unbalanced growth, see Hirschman (1958; 1986); Streeten (1959).

20 Neo-classical 'choice of techniques' argument is exclusively concerned with this aspect. Refer to Shin (1992, section 2.1.2).

21 Dahlman and Westphal (1982: 109–10) summarise the results of empirical studies on the choice of techniques extensively undertaken during the 1960s and 1970s as follows: 'It is now well established that there is a scope for choosing between techniques with different levels of labour productivity, but that scope for choice is by no means uniform.'

22 See Landes (1969: 330) for the German case and Allen (1981a: 8) for the Japanese case.

23 Similarly, strong protective measures taken by the Japanese and the Korean states during their developmental periods may not be an option available to current underdeveloped countries. The GATT system had several clauses supporting protection in developing countries, and the hegemony country during the period, i.e., the US, did not seriously oppose the protective measures in Japan and Korea. However, once the World Trade Organisation (WTO) system, which stipulates few protective measures for developing countries, has been launched, the scope for taking protective measures for current developing countries would be reduced significantly. Even before the launch of the WTO, this trend seemed evident. The so-called second-tier Asian NICs like Thailand, Malaysia, Indonesia, and so on, adopted more liberal policies and were more active in introducing FDI than the first-tier Asian NIC countries.

24 Supple (1976: 307).

25 For instance, Lee (1991: 3) contends that '[i]n a broader sense the overall political environment is a key determinant of the dynamic of economic development'.

26 Of course, different forms of market coordinations can be explained by other factors than the degree of backwardness. For instance, Lundvall (1988) argues that the market tends to be 'organised' if technologies are complex and changing rapidly, and tends to be 'pure' if technologies are simple and changing slowly.

27 In the same vein, Leff (1978: 673) argues that business grouping is an alternative to planned economy.

28 Clapham (1921: 72–3) describes the French system as follows:

> Year by year, under the government of the Restoration, the [protective tariff] system was strengthened and amplified. Iron had started in 1814 with a 50 per cent. duty while steel had only 45 per cent. In 1820 steel got 60 per cent. . . . In 1822 the duty on English iron was raised to 120 per cent. in the interests of the forge-masters and the landowners who supplied them with charcoal. . . . The tariff of 1826 completed the Restoration system. . . . The duties on blankets, cast steel, ropes, and many other commodities were doubled, trebled, quadrupled. . . . The government of Louis Philippe from 1830 to 1848 introduced no fundamental change in tariff policy.

29 Clapham (1921: 236).

30 Clapham (1921: 72). In the same context, Crouzet (1972: 123) argues that,

> [t]his differential customs policy [of Prussia] seems to have been much better conceived than the French tariff system, which prohibited or taxed heavily any kind of imports and encouraged therefore the development of high-cost and inefficient basic industries.

31 Refer to Gerschenkron (1962: 12); Crouzet (1972); Clapham (1921: 71–5); Trebilcock (1981: 147). These contentions, however, need to be scrutinised more carefully, even in the domain of simple economic efficiencies. The resulting patterns of industrial development from the differences in the tariff system are obvious. In Prussia, the later stage iron producers made progress at the expense of traditional pig-iron producers, whereas, in France, many inefficient pig-iron producers could survive at the expense of the later-stage producers. On the other hand, modern coke-smelting showed some progress in France while it was retarded in Germany until the 1840s. The production of coke (or mixed) smelted iron in France rose from 5,000 tons in 1825 to 31,000 tons in 1830, 49,000 tons in 1835, and 82,000 tons in 1840. In contrast, it was only 9,000 tons in 1837 in Prussia (Landes 1969: 217, table 6). In order to calculate the net cost (or benefit) of the French tariff system compared to that of the German system, those comparative gains and losses should be assessed. But a detailed comparative study on this issue was not available to our knowledge.

32 One reason for 'liberal' tariff system would be related with *Junkers'* export interests. But it needs to be answered why the Prussian pig-iron sector could not effectively resist the 'differential' system.

33 Palmer, quoted in Trebilcock (1981: 137).

34 Clapham (1921: 71–2); Trebilcock (1981: 147).

35 In the same vein, Khan (1995) proposes the analysis of 'transition cost' which is the cost of political transition in particular countries as compared to transaction cost which is a maintenance cost of given institutions in a static situation. Trebilcock (1981: 87) also comments that 'the nature and utility of state intervention depended, before anything else, upon the balance of forces within the body politic'.

36 Gunnarsson (1993: 175); Hong (1993: 188).

37 Thus, 'from the 1950s to the 1970s Taiwan had one of the largest public enterprise sectors outside the Communist bloc and sub-Saharan Africa' (Whitley 1992: 58).

38 Young-Chul Park (1987); Whitley (1992: 58).

39 Hou and Gee (1993: 407); Table 8.4 in section 8.3.

40 Mody (1990: 299).

41 The initial beginning of ERSO was similar to that of KIET in Korea since both

were set up as supporting institutions for its private sector. But, in contrast to the case of ERSO, the role of KIET was just *transitory*. It transferred its R&D facilities to the private sector in the 1980s as *chaebols* began to play a leading role in R&D investment (section 8.3).

42 Hou and Gee (1993: 396–9); Mody (1990: 299).

43 For instance, Mody (1990) argues that Korea is in a better position because technological trend in the electronics industry is towards increasing economies of scale and scope.

44 Emphasis added.

45 Gerschenkron (1962: 43) also argues that,

> [t]here is not a precise system of weights by virtue of which disparate factors could be brought together over a common denominator; nor could we possibly determine the precise quantities of the pertinent factors to which those weights could be applied.

# BIBLIOGRAPHY


Abe, E. (1991) 'Introduction', in Abe, E. and Suzuki, Y. (1991) *Changing Patterns of International Rivalry: Some Lessons from the Steel Industry*, Tokyo: University of Tokyo Press.

Abramovitz, M. (1979) 'Rapid Growth Potential and Its Realization: The Experience of Capitalist Economies in the Postwar Period', in Malinvaud, E. (ed.) (1979) *Economic Growth and Resources, vol. 1, The Major Issues*, London: Macmillan; in Abramovitz, M. (1989) *Thinking About Growth*, Cambridge: Cambridge University Press.

Abramovitz, M. (1986) 'Catching-Up, Forging Ahead and Falling Behind', *Journal of Economic History* 46/2: 385–406; in Abramovitz, M. (1989) *Thinking About Growth*, Cambridge: Cambridge University Press.

Abramovitz, M. (1989) 'Thinking About Growth', in Abramovitz, M. (1989) *Thinking About Growth*, Cambridge: Cambridge University Press.

Allen, G.C. (1940) 'Japanese Industry: Its Organization and Development to 1937', in Schumpeter, E.B. (ed.) (1940) *The Industrialization of Japan and Manchukuo*, New York: Macmillan.

Allen, G.C. (1965) *Japan's Economic Expansion*, Oxford: Oxford University Press.

Allen, G.C. (1981a) *A Short Economic History of Modern Japan*, 4th edn, Macmillan: London.

Allen, G.C. (1981b) 'Industrial Policy and Innovation in Japan', in Carter, C. (ed.) (1981) *Industrial Policy and Innovation*, London: Heinemann.

Allen, R.C. (1979) 'International Competition in Iron and Steel, 1850–1913', *Journal of Economic History* 39/4: 911–37.

Ames, E and Rosenberg, N. (1963) 'Changing Technological Leadership and Industrial Growth', *Economic Journal*, March.

Amsden, A.H. (1989) *Asia's Next Giant*, New York: Oxford University Press.

Amsden, A.H. (1991) 'Diffusion of Development: The Late-Industrialising Model and Greater East Asia', *American Economic Review* 81/2: 282–6.

Archambault, E.J. (1991) 'Small is Beautiful, Large is Powerful: Manufacturing Semiconductors in South Korea', unpublished M.Sc. dissertation, Science Policy Research Unit, University of Sussex.

Arrow, K. (1962) 'Economic Implications of Learning by Doing', *Review of Economic Studies* 29.

Barro, R.J. (1991) 'Economic Growth in a Cross Section of Countries', *Quarterly Journal of Economics*, May.

Baumol, W.J. (1986), 'Productivity Growth, Convergence, and Welfare: What the Long-Run Data Show', *American Economic Review* 76/5.

Bell, M. and Pavitt, K. (1993) 'Accumulating Technological Capability in Developing

Countries', Proceedings of the World Bank Annual Conference on Development Economics.

Black, C.E. *et al.* (1975) *The Modernization of Japan and Russia: A Comparative Study*, London: Free Press.

Boadway, R.W. and Bruce, N. (1984) *Welfare Economics*, Oxford: Blackwell.

Bohm, P. (1987) 'Second Best', in Eatwell, J. *et al.* (eds) (1987) *The New Palgrave: A Dictionary of Economics*, Vol. 4, London: Macmillan.

Böhme, H. (1978) *An Introduction to the Social and Economic History of Germany*, London.

Borchardt, K. (1973) 'Germany 1700–1914', in Cipolla, C.M. (ed.) (1973) *The Fontana Economic History of Europe: 4 The Emergence of Industrial Societies*, Vol. 1, London and New York: Harvester.

Borrus, M., Millstein, J.E. and Zysman, J. (1983) 'Trade and Development in the Semiconductor Industry: Japanese Challenge and American Response', in Zysman, J. and Tyson, L. (eds) (1983) *American Industry in International Competition*, Ithaca and London: Cornell University Press.

Boulding, K.E. (1957) 'A New Look at Institutionalism', *American Economic Review* 47/2: 1–12.

Boyer, R. (1988) 'Technical Change and the Theory of "Régulation"', in Dosi, G. *et al.*, (1988) *Technical Change and Economic Theory*, London and New York: Pinter.

Broadberry, S. (1992) 'Manufacturing and the Convergence Hypothesis: What the Long Run Data Show', Warwick Economic Research Papers, no. 393, University of Warwick.

Broadberry, S. (1993) 'Technological Leadership and Productivity Leadership in Manufacturing since the Industrial Revolution: Implications for the Convergence Debate', Warwick Economic Research Papers, no. 414, University of Warwick.

Byun, H.Y. (1980) 'Hankuk ChulKang Kongup eui Kisul Chujuck: Pohang Jechul eul Jungsim euro' (Technological Accumulation in the Korean Iron and Steel Industry: With Special Reference to POSCO), Kyungje Nonjip (Seoul National University) 19/2: 124–36.

Cameron, R.E. (1956) 'Founding the Bank of Darmstadt', *Explorations in Entrepreneurial History* 9.

Chandler, A. Jr. (1990) *Scale and Scope*, Cambridge, Mass.: Harvard University Press.

Chang, H.J. (1990) 'Interpreting the Korean Experience: Heaven or Hell', Research Paper Series, no. 42, mimeo, Faculty of Economics, University of Cambridge.

Chang, H.J. (1994) *The Political Economy of Industrial Policy*, London: Macmillan.

Cho, D.S. (1991) *Hankook Chaebol Yeonkoo* (A Study of Korean Chaebol), *Maeil Kyungje–The Economic Daily*: Seoul.

Cho, M.H. (1995) *Bandoche Sanup* (The Semiconductor Industry), Seoul: Woongjin.

Choi, Young Rak (1991) *Bandoche Gisul Baljun eul wihan Jache Gisul Neungryuk Chukjuk ei gwanhan Yungoo* (A Study on Accumulation of Indigenous Technological Capability for the Development of the Semiconductor Industry), Seoul: Korea Institute for Science and Technology Policy.

Choi, Young Rak (1994) 'Dynamic Techno-Management Capability', unpublished Ph.D. thesis, Department of Economics and Planning, Roskilde University.

Chu, D.Y. (1992) *Bandoche Sanup eui Geupgyukhan Byunhwa wa Uri eui Daeung* (Rapid Changes in the Semiconductor Industry and Our Responses), Seoul: KIET.

Clapham, J.H. (1921) *The Economic Development of France and Germany 1815–1914*, Cambridge: Cambridge University Press.

Cole, D.C. and Lyman, P.N. (1971) *Korean Development: The Interplay of Politics and Economics*, Cambridge, Mass.: Harvard University Press.

Collins, R. (1986) *Weberian Sociological Theory*, Cambridge: Cambridge University Press.

Commons, J.R. (1934) *Institutional Economics: Its Place in Political Economy*, New York: Macmillan.

Crouzet, F. (1972) 'Western Europe and Great Britain: "Catching-Up" in the First Half of the Nineteeth Century', in Youngson, A.J. (ed.) (1972) *Economic Development in the Long Run*, London: George Allen & Unwin.

Cumings, B. (1987) 'The Origins of Development of the Northeast Asian Political Economy: Industrial Sectors, Product Cycles, and Political Consequences', in Deyo, F.C. (ed.) (1987) *The Political Economy of the New Asian Industrialism*, Ithaca, N.Y. and London: Cornell University Press.

Dahlman, C. and Frischtak, C.R. (1993) 'National Systems Supporting Technical Advance in Industry: The Brazilian Experience', in Nelson, R. (ed.) (1993) *National Innovation Systems: A Comparative Analysis*, Oxford: Oxford University Press.

Dahlman, C. and Nelson, R. (1993) 'Social Absorption Capability, National Innovation Systems and Economic Development', paper prepared for UNU/INTECH conference, Maastricht, The Netherlands, June.

Dahlman, C.J and Westphal, L. (1982) 'Technological Effort in Industrial Development: A Survey', in Stewart, F. and James, J. (eds) (1982) *The Economics of New Technology in Developing Countries*, London: Frances Pinter.

Datta-Chauduri, M.K. (1981) 'Industrialisation and Foreign Trade: The Development Experience of South Korea and the Philippines', in Lee, E. (ed.) (1981), *Export-Led Industrialisation and Development*, Geneva: ILO.

De Long, J.B. and Summers, L. (1991) 'Equipment Investment and Economic Growth', *Quarterly Journal of Economics* 106: 445–502.

Denison, E.F. (1964) 'The Unimportance of the Embodied Question', *American Economic Review* 54/2: 90–4.

Deyo, F.C. (1987) 'Introduction', in Deyo, F.C. (ed.) (1987) *The Political Economy of the New Asian Industrialism*, Ithaca, N.Y. and London: Cornell University Press.

Diefendorf, J.M. (1980) *Businessmen and Politics in the Rheinland*, Princeton, N.J.: Princeton University Press.

Dore, R. (1986) *Flexible Rigidities: Industrial Policy and Structural Adjustment in the Japanese Economy 1970–80*, London: Athlone Press.

Dorfman, J. *et al.* (1963) *Institutional Economics: Veblen, Commons, and Mitchell Reconsidered*, Berkeley, Calif.: University of California Press.

Dosi, G. (1981) 'Institutions and Markets in High Technology: Government Support for Micro-Electronics in Europe', in Carter, C. (ed.) (1981) *Industrial Policy and Innovation*, London: Heinemann.

Dosi, G. (1984) *Technical Change and Industrial Transformation: The Theory and Application to the Semiconductor Industry*, London: Macmillan.

Dosi, G. and Orsenigo, L. (1988) 'Coordination and Transformation: An Overview of Structures, Behaviours and Change in Evolutionary Environments', in Dosi, G. *et al.* (eds) (1988) *Technical Change and Economic Theory*, London and New York: Pinter.

Dowrick, S. (1991) 'Technological Catch-Up and Diverging Incomes: Patterns of Economic Growth 1960–1988', *Economic Journal* 102/412: 600–10.

Dowrick, S. and Gemmell, N. (1991) 'Industrialisation, Catching Up and Economic Growth: A Comparative Study Across the World's Capitalist Economies', *Economic Journal* 101/405: 263–75.

Elbaum, B. and Lazonick, W. (eds) (1986) *The Decline of the British Economy*, Oxford: Clarendon Press.

Elsevier Advanced Technology (1993; 1991; 1987) *A Profile of the Worldwide Semiconductor Industry*, London: Elsevier.

Enos, J.L. and Park, W.H. (1988) *The Adoption and Diffusion of Imported Technology: The Case of Korea*, New York: Croom Helm.
Enos, J.L. (1991) *The Creation of Technological Capability in Developing Countries*, London: Pinter.
Ernst, D. (1983) *The Global Race in Microelectronics: Innovation and Corporate Strategies in a Period of Crisis*, Frankfurt: Campus Verlag.
Ernst, D. and O'Connor, D. (1989) *Technology and Global Competition: The Challenge for Newly Industrialising Economies*, Paris: OECD.
Eucken, W. (1950) *The Foundations of Economics: History and Theory in the Analysis of Economic Reality*, trans. by T.W. Hutchison, London.
Feldenkirchen, W. (1981) 'The Banks and the Steel Industry in the Ruhr: Developments in Relation from 1873–1914', in Engels, W. and Pohl, H. (1981) *German Yearbook on Business History 1981*, Berlin and New York: Springer-Verlag.
Fischer, W. (1963) 'Government Activity and Industrialisation in Germany 1815–70', in Rostow, W.W. (ed.) (1963) *The Economics of Take-off into Sustained Growth*, London.
Fischer, W. (1968) 'Gustav Schmoller', in Sills, D. (ed.) (1968) *International Encyclopedia of the Social Sciences* Vol. 14.
Flaherty, M.T. and Itami, H. (1984) 'Finance', in Okimoto, D. *et al.* (eds) (1984) *Competitive Edge: The Semiconductor Industry in the US and Japan*, Stanford, Calif.: Stanford University Press.
Fransman, M. (1984) 'Introduction', in Fransman, M. and King, K. (eds) (1984) *Technological Capability in the Third World*, London: Macmillan.
Fransman, M. (1985) 'Conceptualising Technical Change in the Third World in the 80s: An Interpretive Survey', *Journal of Development Studies* 21/4.
Fransman, M. (1990) *The Market and Beyond: Cooperation and Competition in Information Technology in the Japanese System*, Cambridge: Cambridge University Press.
Freeman, C. (1987) *Technology Policy and Economic Performance: Lessons from Japan*, London: Pinter.
Freeman, C. (1988) 'Japan: A New National Innovation System?', in Dosi, G. *et al.* (eds) (1988) *Technical Change and Economic Theory*, London and New York: Pinter.
Freeman, C. (1989) 'New Technology and Catching-up', in Cooper, C. and Kaplinsky, R. (eds) (1989) *Technology and Development in the Third Industrial Revolution*, London: Frank Cass.
Freeman, C. (1992) 'The Nature of Innovation and the Evolution of the Productive System', in Freeman, C. (1992) *The Economics of Hope: Essays on Technical Change, Economic Growth and the Environment*, London: Pinter.
Freeman, C., Clark, J. and Soete, L. (1982) *Unemployment and Technical Innovation*, Brighton: University of Sussex Press.
Freeman, C. and Perez, C. (1988) 'Structural Crises of Adjustment: Business Cycles and Investment Behaviour', in Dosi, G. *et al.* (eds) (1988) *Technical Change and Economic Theory*, London and New York: Pinter.
Fremdling, R. (1977) 'Railroads and German Economic Growth: A Leading Sector Analysis with a Comparison to the United States and Great Britain', *Journal of Economic History* 37.
Fremdling, R. (1991a) 'Foreign Competition and Technological Change: British Exports and the Modernisation of the German Iron Industry from the 1820s to the 1860s', in Lee, W.R. (ed.) (1991) *German Industry and German Industrialisation: Essays in German Economic and Business History in the Nineteenth and Twentieth Century*, London and New York: Routledge.

Fremdling, R. (1991b) 'The German Iron and Steel Industry in the 19th Century', in Abe, E. and Suzuki, Y. (1991) *Changing Patterns of International Rivalry: Some Lessons from the Steel Industry*, Tokyo: University of Tokyo Press.

Gale, W.K.V. (1990) 'Ferrous Metals', in McNeil, I. (ed.) (1990) *An Encyclopedia of the History of Technology*, London and New York: Routledge.

Gay, E.F. (1954) 'The Tasks of Economic History', in Lane, F.C. and Riemersma, J.C. (eds) (1954) *Enterprise and Secular Change: Readings in Economic History*, London: George Allen & Unwin.

Gerschenkron, A. (1962) *Economic Backwardness in Historical Perspective*, Cambridge, Mass.: Harvard University Press.

Gerschenkron, A. (1963) 'The Early Phases of Industrialization in Russia: Afterthoughts and Counterthoughts', in Rostow, W.W. (ed.) (1963) *The Economics of Take-Off into Sustained Growth*, London: Macmillan.

Gerschenkron, A. (1968) *Continuity in History and Other Essays*, Cambridge, Mass.: Harvard University Press.

Gerschenkron, A. (1970) *Europe in the Russian Mirror, Four Lectures in Economic History*, Cambridge: Cambridge University Press.

Giddens, A. (1971) *Capitalism and Modern Social Theory: An Analysis of the Writings of Marx, Durkheim and Max Weber*, Cambridge: Cambridge University Press.

Gregory, G. (1986) *Japanese Electronics Technology: Enterprise and Innovation*, New York: John Wiley & Sons.

Gunnarsson, C. (1993) 'Dirigism or Free-Trade Regime? A Historical and Institutional Interpretation of the Taiwanese Success Story', in Hansson, G. (ed.) (1993) *Trade, Growth and Development: The Role of Politics and Institutions*, London: Routledge.

Henderson, J. (1989) *The Globalisation of High Technology Production*, London: Routledge.

Henderson, W.O. (1967) *The State and the Industrial Revolution in Prussia 1740–1870*, Liverpool: Liverpool University Press.

Henderson, W.O. (1975) *The Rise of German Industrial Power: 1834–1914*, London: Temple Smith.

Henderson, W.O. (1984) *The Zollverein*, London: Frank Cass.

Henderson, W.O. (1985) *The Industrial Revolution on the Continent: Germany, France, Russia 1800–1914*, London: Frank Cass.

Hirsch, S. (1965) 'The US Electronics Industry and International Trade', *National Institute Economic Review* 34: 92–4.

Hirschman, A.O. (1958) *The Strategy of Economic Development*, New Haven, Conn.: Yale University Press.

Hirschman, A.O. (1968) 'The Political Economy of Import-Substituting Industrialization in Latin America', reprinted in Hirschman, A.O. (1985) *Bias for Hope*, London: Westview Press.

Hirschman, A.O. (1986) 'A Dissenter's Confession: The Strategy of Economic Development Revisited', in *Rival Views of Market Society*, New York: Viking.

Hirschmeier, J. (1964) *The Origins of Entrepreneurship in Meiji Japan*, Cambridge, Mass.: Harvard University Press.

Hobday, M. (1989) 'Corporate Strategies in the International Semiconductor Industry', *Research Policy* 18: 225–38.

Hobday, M. (1991) 'Semiconductor Technology and the Newly Industrializing Countries: The Diffusion of ASICs (Application Specific Integrated Circuits)', *World Development* 19/4: 375–97.

Hobday, M. (1992) 'The European Electronics Industry: Technological and Structural Change', *Technovation* 12/2: 75–97.

Hobday, M. (1994) 'How the Four Asian Dragons Caught Up in Electronics: Strategies of Latecomer Firms', forthcoming in *World Development*.

Hobday, M. (1995) *Innovation in East Asia: The Challenge to Japan*, London: Edward Elgar.

Hodgson, G. (1991) 'Institutional Economic Theory: the Old versus the New', in Hodgson, G. '1991', *After Marx and Sraffa: Essays in Political Economy*, London: Macmillan.

Hogan, W.T. (1983) *World Steel in the 1980s: A Case of Survival*, Toronto: Lexington Books.

Hogan, W.T. (1990) 'South Korean Steel Growth Spurs Industrial Development', *Iron and Steel Engineer*, November.

Hong, W.T. (1993) 'Trade and Development: The Experience of Korea and Taiwan', in Hansson, G. (ed.) (1993) *Trade, Growth and Development: The Role of Politics and Institution*, London: Routledge.

Hou, C.M. and Gee, S. (1993) 'National Systems Supporting Technical Advance in Industry: The Case of Taiwan', in Nelson, R. (ed.) (1993) *National Innovation Systems: A Comparative Analysis*, Oxford: Oxford University Press.

Jacobsson, S. (1993) 'The Length of the Learning Period: Evidence from the Korean Engineering Industry', *World Development* 21/3.

Johnson, C. (1982) *MITI and the Japanese Miracle: The Growth of Industrial Policy 1925–1975*, Stanford, Calif.: Stanford University Press.

Johnson, C. (1987) 'Political Institutions and Economic Performance: The Government–Business Relationship in Japan, South Korea, and Taiwan', in Deyo, F.C. (ed.) (1987) *The Political Economy of the New Asian Industrialism*, Ithaca, N.Y. and London: Cornell University Press.

Jones, L.P. and Sakong, I. (1980) *Government, Business, and Entrepreneurship in Economic Development: The Korean Case*, Cambridge, Mass.: Harvard University Press.

Juhn, S. (1991) 'Challenge of a Latecomer: The Case of the Korean Steel Industry with Specific Reference to POSCO', in Abe, E. and Suzuki, Y. (1991) *Changing Patterns of International Rivalry: Some Lessons from the Steel Industry*, Tokyo: University of Tokyo Press.

Kahan, A. (1966) 'Social Structure, Public Policy, and the Development of Education and the Economy in Czarist Russia', in Anderson, C.A. and Bowman, M.J. (eds) (1966) *Education and Development*, London: Frank Cass.

Kaldor, N. (1957) 'A Model of Economic Growth', *Economic Journal* 67: 591–624.

Kaldor, N. (1961) 'Capital Accumulation and Economic Growth', in Targetti, F. and Thirlwall, A.P. (eds) (1989) *The Essential Kaldor*, London: Duckworth.

Kaldor, N. (1966) 'Causes of the Slow Rate of Economic Growth in the United Kingdom', in Targetti, F. and Thirlwall, A.P. (eds) (1989) *The Essential Kaldor*, London: Duckworth.

Kaldor, N. (1975) 'Economic Growth and The Verdoorn Law: A Comment on Mr Rowthorn's Article', *Economic Journal* 85: 891–6.

Kaplinsky, R. (1989) 'Technological Revolution and the International Division of Labour in Manufacturing: A Place for the Third World?', in Cooper, C. and R. Kaplinsky (eds) (1989) *Technology and Development in the Third Industrial Revolution*, London: Frank Cass.

KEIPA (Korea Electronics Industry Promotion Association) (1988) *Bandoche Sanup eui Hyunhwang gwa Junmang* (Current Situation and Prospect of the Semiconductor Industry), Seoul: KEIPA.

Kennedy, P. (1989) *The Rise and Fall of the Great Powers: Economic Change and Military Conflict from 1500 to 2000*, London: Fonta Press.

Khan, M.H. (1989) 'Clientelism, Corruption and Capitalist Development', unpublished Ph.D. thesis, University of Cambridge, forthcoming Oxford University Press.

Khan, M.H. (1995) 'State Failure in Weak State: A Critique of New Institutionalist Explanations', forthcoming in Harris, J., Hunter, H. and Lewis, C. (eds) *New Institutional Economics and Third World Development*, London: Routledge.

Kim, Hwan Suk, Hong, Sung Bum and Lee, Young Hee (1992) *Segye Gyungje eui Jangki Padong gwa Shinkisul eui Kukje Hwaksan: Jangki Padong Iron eui Hankook eieui Shisajum* (Long Waves in the World Economy and Diffusion of New Technologies: Implications of Long Wave Theories to Korea), Seoul: Korea Institute for Science and Technology Policy.

Kim, J.H. (1990) 'Korean Industrial Policy in the 1970s: The Heavy and Chemical Industry Drive', KDI Working Paper, no. 9015, Seoul: KDI.

Kim, Linsu (1980) 'Stages of Development of Industrial Technology in a LDC: A Model', *Research Policy* 9/3: 254–77.

Kim, Linsu (1991), 'Pros and Cons of International Technology Transfer: A Developing Country View', in Agmon, T. and von Glinow, M.A. (eds) (1991) *Technology Transfer in International Business*, Oxford: Oxford University Press.

Kim, Linsu (1993) 'National System of Industrial Innovation: Dynamics of Capability Building in Korea', in Nelson, R. (ed.) (1993) *National Innovation Systems: A Comparative Analysis*, Oxford: Oxford University Press.

Kim, Linsu and Dahlman, C.J. (1992) 'Technology Policy for Industrialisation: An Integrative Framework and Korea's Experience', *Research Policy* 21/5: 437–52.

Kim, Linsu and Lee, Jinjoo (1987) 'Korea's Entry into the Computer Industry and Its Acquisition of Technological Capability', *Technovation*, Sept.: 277–93.

Kim, Yong Bok (1991) 'Junja Sanup eui Jabon Chujuk Gwajung gwa Oigukin Tuja' (Capital Accumulation in the Electronics Industry and Foreign Investment), in Yang, Woo-Jin *et al.* (1991) *Hankuk Jabonjuei eui Bunseok* (An Analysis of Korean Capitalism), Seoul: Ilbit.

Kiyonari, T. and Nakamura, H. (1980) 'The Establishment of the Big Business System', in Sato, K. (ed.) (1980) *Industry and Business in Japan*, White Plains, N.Y.: Sharpe.

Koh, Dong Jin (1992) 'Beyond Technological Dependency, Towards an Agile Giant: The Strategic Concerns of Korea's Samsung Electronics Co. for the 1990s', unpublished M.Sc. dissertation, Science Policy Research Unit, University of Sussex.

Kravis, I. (1970) 'Trade as a Handmaiden of Growth: Similarities between the Nineteenth and Twentieth Century', *Economic Journal*, Dec.

Kuisel, R.F. (1981) *Capitalism and the State in Modern France*, Cambridge: Cambridge University Press.

Lall, S. (1987) *Learning to Industrialise: The Acquisition of Technological Capability in India*, London: Macmillan.

Landes, D. (1965) 'Japan and Europe: Contrasts in Industrialization', in Lockwood, W.W. (ed.) (1965) *The State and Economic Enterprise in Japan*, Princeton, N.J.: Princeton University Press.

Landes, D. (1969) *The Unbound Prometheus: Technological Change and Industrial Development in Western Europe from 1750 to the Present*, Cambridge: Cambridge University Press.

Lane, F.C. and Riemersma, J.C. (1954) 'Introduction to Arthur Spiethoff', in Lane, F.C. and Riemersma, J.C. (eds) (1954) *Enterprise and Secular Change: Readings in Economic History*, London: George Allen & Unwin.

Langlois, R.N. (1986) 'The New Institutional Economics: An Introductory Essay', in Langlois, R.N. (ed.) (1986) *Economics as a Process: Essays in the New Institutional Economics*, Cambridge: Cambridge University Press.

Lee, W.R. (1991) 'The Paradigm of German Industrialisation: Some Recent Issues and Debates in the Modern Historiography of German Industrial Development', in Lee, W.R. (ed.) (1991) *German Industry and German Industrialisation: Essays in German Economic and Business History in the Nineteenth and Twentieth Century*, London and New York: Routledge.

Leff, N.H. (1978) 'Industrial Organization and Entrepreneurship in the Developing Countries: The Economic Groups', *Economic Development and Cultural Change*, July.

Lockwood, W.W. (1954) *The Economic Development of Japan*, Princeton, N.J.: Princeton University Press.

Lucas, R.E. Jr. (1988) 'On the Mechanics of Economic Development', *Journal of Monetary Economics* 22.

Lundvall, B-A. (1985) *Product Innovation and User-Producer Interaction*, Aalborg.

Lundvall, B-A. (1988) 'Innovation as an Interactive Process: From User-Producer Interaction to the National Systems of Innovation', in Dosi, G. *et al.* (eds) (1988) *Technical Change and Economic Theory*, London and New York: Pinter.

Lundvall, B-A. (ed.) (1992) *National Systems of Innovation: Towards a Theory of Innovation and Interactive Learning*, London: Pinter.

Lynn, L. (1982) *How Japan Innovates: A Comparison with the US in the Case of Oxygen Steel-making*, Boulder, Col.

Lynn, L.H. (1985) 'Technology Transfer to Japan: What We Know, What We Need to Know, and What We Know That May Not be So', in Rosenberg, N. and Frischtak, C. (eds) (1985) *International Technology Transfer: Concepts, Measures and Comparisons*, New York: Praeger.

McCloskey, D.N. (1991) 'Kinks, Tools, Spurts and Substitutes: Gerschenkron's Rhetoric of Relative Backwardness', in Sylla, R. and Tonilo G. (eds) (1991) *Patterns of European Industrialization: The Nineteenth Century*, London: Routledge.

McCraw, T. and O'Brien, P. (1986) 'Production and Distribution', McCraw, T. (ed.) (1986) *America versus Japan*, Boston, Mass.: Harvard Business School Press.

Mackintosh Consultants (1982) *A Profile of the Worldwide Semiconductor Industry*, Luton: Mackintosh.

Maddison, A. (1982) *Phases of Capitalist Development*, Oxford: Oxford University Press.

Maddison, A. (1991) *Dynamic Forces in Capitalist Development: A Long-Run Comparative View*, Oxford: Oxford University Press.

*Maeil Kyungje–The Economic Daily* (Korean), several issues.

*The Maekyung Business Week* (Korean), several issues.

Mason, E.S. *et al.* (1980) *The Economic and Social Modernization of the Republic of Korea*, Cambridge, Mass.: Harvard University Press.

Matthews, R.C.O. (1986) 'The Economics of Institutions and the Sources of Economic Growth', *Economic Journal*.

Meinecke, F. (1977) *The Age of German Liberation 1795–1815*, Berkeley, Calif. and London: University of California Press.

Methé, D.T. (1992) 'The Influence of Technology and Demand Factors on Firm Size and Industrial Structure in the DRAM market 1973–1988', *Research Policy* 21: 13–25.

Millstein, J.E. (1983) 'Decline in an Expanding Industry: Japanese Competition in Color Television', in Zysman, J. and Tyson, L. (eds) (1983) *American Industry in International Competition*, Ithaca, N.Y. and London: Cornell University Press.

Miyashita, K. and Russell (1994) *Keiretsu: Inside the Hidden Japanese Conglomerates*, Tokyo: McGraw-Hill.

Miyazaki, Y. (1980a), 'The Japanese-Type Structure of Big Business', in Sato, K. (ed.) (1980) *Industry and Business in Japan*, White Plains, N.Y.: Sharpe.

Miyazaki, Y. (1980b) 'Excessive Competition and the Formation of Keiretsu', in Sato, K. (ed.) (1980) *Industry and Business in Japan*, White Plains, N.Y.: Sharpe.

Mody, A. (1990) 'Institutions and Dynamic Comparative Advantage: the Electronics Industry in South Korea and Taiwan', *Cambridge Journal of Economics* 14: 291–314.

Morikawa, H. (1982) 'The Zaibatsu in the Japanese Iron and Steel Industry', in Pöhl, H. (ed.) (1982) *Innovation, Know How, Rationalization and Investment in the German and Japanese Economies 1868/1871–1930/1980*, Wiesbaden: Franz Steiner Verlag GmbH.

Morishima, M. (1982) *Why Has Japan 'Succeeded'? Western Technology and the Japanese Ethos*, Cambridge: Cambridge University Press.

MOST (Ministry of Science and Technology) (1993) *Kwahak Kisul Yunku Hwaldong Chosa Bogo* (Report on the Survey of Research and Development in Science and Technology), Seoul: MOST.

Nagashima, O. (1991) 'Comments on Okazaki', in Abe, E. and Suzuki, Y. (1991) *Changing Patterns of International Rivalry: Some Lessons from the Steel Industry*, Tokyo: University of Tokyo Press.

Nelson, R.R. (1964) 'Aggregate Production Function and Medium-Range Growth Projection', *American Economic Journal*, Sept.

Nelson, R.R. (1988) 'Institutions Supporting Technical Change in the United States', in Dosi, G. *et al.* (1988) *Technical Change and Economic Theory*, London and New York: Pinter.

Nelson, R.R. (1990) 'Capitalism as an Engine of Progress', *Research Policy* 19: 193–214.

Nelson, R.R (ed.) (1993) *National Innovation Systems: A Comparative Analysis*, Oxford: Oxford University Press.

Nielsen, K. (1991) 'Towards a Flexible Future: Theories and Politics', in Jessop *et al.* (eds) (1991) *The Politics of Flexibility: Restructuring State and Industry in Britain, Germany and Scandinavia*, Basingstoke: Edward Elgar.

Nolan, J. E. (1986) *Military Industry in Taiwan and South Korea*, New York: St. Martin's Press.

North, D. (1981) *Structure and Change in Economic History*, New York: Norton.

North, D. (1989) 'Institutions and Economic Growth: An Historical Introduction', *World Development* 17/9: 1319–32.

North, D. (1990) *Institutions, Institutional Change and Economic Performance*, Cambridge: Cambridge University Press.

Nurkse, R. (1953) *Problems of Capital Formation in Underdeveloped Countries*, Oxford: Oxford University Press.

Ohkawa, K. and Rosovsky, H. (1973) *Japanese Economic Development: Trend Acceleration in the Twentieth Century*, Stanford, Calif.: Stanford University Press.

Okazaki, T. (1991) 'Import-substitution and Competitiveness in the Prewar Japanese Iron and Steel Industry', in Abe, E. and Suzuki, Y. (eds) (1991) *Changing Patterns of International Rivalry: Some Lessons from the Steel Industry*, Tokyo: University of Tokyo Press.

Okichi, A. and Yasuoka, S. (ed.) (1984) *Family Business in the Era of Industrial Growth: Its Ownership and Management*, Tokyo: University of Tokyo Press.

Okimoto, D. (1984) 'Political Context', in Okimoto, D. *et al.* (eds) (1984) *Competitive Edge: The Semiconductor Industry in the USA and Japan*, Stanford, Calif.: Stanford University Press.

Okuno-Fujiwara, M. (1991) 'Industrial Policy in Japan: A Political Economy View', in

Krugman, P. (ed.) (1991) *Trade with Japan: Has the Door Opened Wider?*, Chicago, Ill.: Chicago University Press.

Pack, H. (1982) 'Aggregate Implications of Factor Substitution in Industrial Processes', *Journal of Development Economics* 11: 1–37.

Park, Hee Chun *et al.* (1987) *Uri Nara Bandoche Sanup eui Hyunhwang gwa Yunksung Junryak* (The Current Situation of the Korean Semiconductor Industry and a Strategy for Its Promotion), Seoul: KIET.

Park, Yung Chul (1986) 'Foreign Debt, Balance of Payments, and Growth Prospects: The Case of the Republic of Korea', *World Development* 14/8: 1019–58.

Park, Yung Chul (1987) 'Development Lessons from Asia: The Role of Government in South Korea and Taiwan', *American Economic Review* 80/2: 118–21.

Park, Yong Chan (1987) 'The National System of Innovation in Korea with an Introduction to the Semiconductor Industry', unpublished M.Sc. dissertation, Science Policy Research Unit, University of Sussex.

Paskoff, P.F. (1991) 'The Growth of the American Steel Industry 1865–1914: Technological Change, Capital Investment, and Trade Policy', in Abe, E. and Suzuki, Y. (eds) (1991) *Changing Patterns of International Rivalry: Some Lessons from the Steel Industry*, Tokyo: University of Tokyo Press.

Patrick, H.T. and Rosovsky, H. (eds) (1976) *Asia's Next Giant*, Washington, D.C.: Brookings Institute.

Perez, C. (1983) 'Structural Change and the Assimilation of New Technologies in the Economic and Social System', *Futures* 15/4: 357–75.

Perez, C. (1985) 'Micro-Electronics, Long Waves and World Structural Change: New Perspectives for Developing Countries', *World Development* 13/3: 441–63.

Perez, C. and Soete, L. (1988) 'Catching-up in Technology', in Dosi, G. *et al.* (eds) (1988) *Technical Change and Economic Theory*, London and New York: Pinter.

Pollard, S. (1973) 'Industrialization and the European Economy', *Economic History Review* 26.

Pollard, S. (1981) *Peaceful Conquest: The Industrialization of Europe 1760–1970*, Oxford: Oxford University Press.

Popper, K. (1960) *The Poverty of Historicism*, London: Routledge.

POSCO (1989) *Pochul Isipnyun Sa* (20 Years' History of POSCO), Pohang.

Pounds, N.J.G. (1952) *The Ruhr: A Study in Historical and Economic Geography*, London: Faber & Faber.

Pounds, N.J.G. (1958) *The Upper Silesian Industrial Region*, Bloomington, Ind.: Indiana University Publications.

Pounds, N.J.G. (1959) 'Economic Growth of Germany', in Aitken, H.G.J. (ed.) (1959) *The State and Economic Growth*, New York.

Pounds, N.J.G. and Parker, W.N. (1957) *Coal and Steel in Western Europe*, London.

Reddy, N.M. and Zhao, L. (1990) 'International Techonology Transfer: A Review', *Research Policy* 19.

Rhee, Y.W., Ross-Larson, B. and Pursell, G. (1984) *Korea's Competitive Edge: Managing the Entry into World Market*, Baltimore, Md. and London: The Johns Hopkins University Press.

Riedel, J. (1984) 'Trade as the Engine of Growth in Developing Countries, Revisited', *Economic Journal* 90: 56–73.

Romer, P. (1986) 'Increasing Returns and Long-Run Growth', *Journal of Political Economy* 94/5.

Romer, P. (1989) 'Capital Accumulation in the Theory of Long-Run Growth', in Barro (ed.) (1989) *Modern Business Cycle Theory*, Oxford: Basil Blackwell.

Rosenberg, N. (1976) *Perspectives on Technology*, Cambridge: Cambridge University Press.

Rosenberg, N. (1982) *Inside the Blackbox*, Cambridge: Cambridge University Press.

Rosenstein-Rodan, P.N. (1943) 'Problems of Industrialization of Eastern and South Eastern Europe', *Economic Journal* 53.

Rosovsky, H. (1961) *Capital Formation in Japan 1868–1940*, New York.

Rosovsky, H. (1965) 'The Take-Off into Sustained Controversy', *Journal of Economic History* 25/2: 271–5.

Rosovsky, H. (1972) 'What are the "Lessons" of Japanese Economic History?' in Youngson, A.J. (ed.) (1972) *Economic Development in the Long Run*, London: George Allen & Unwin.

Rostow, W.W. (ed.) (1963) *The Economics of Take-Off into Sustained Growth*, London: Macmillan.

Rostow, W.W. (1971) *The Stages of Economic Growth: A Non-Communist Manifesto*, 2nd edn, Cambridge: Cambridge University Press.

Rowthorn, R.E. (1975a), 'What Remains of Kaldor's Law?' *Economic Journal* 83: 10–19.

Rowthorn, R.E. (1975b), 'A Reply to Lord Kaldor's Comment', *Economic Journal* 85: 897–901.

Rowthorn, R.E. and Wells, J.R. (1987) *De-Industrialisation and Foreign Trade*, Cambridge: Cambridge University Press.

Runciman, W.G. (ed.) (1978) *Weber: Selections in Translation*, Cambridge: Cambridge University Press.

Schumpeter, J. (1934) *The Theory of Economic Development*, Cambridge, Mass.: Harvard University Press.

Schumpeter, J. (1939) *Business Cycles I–II*, New York: McGraw-Hill.

Schumpeter, E.B. (1940) 'Industrial Development and Government Policy 1936–1940', in Schumpeter, E.B. (ed.) (1940) *The Industrialization of Japan and Manchukuo*, New York: Macmillan.

Schumpeter, J. (1943) *Capitalism, Socialism and Democracy*, New York: Harper.

Schumpeter, J. (1954) *History of Economic Analysis*, London: George Allen & Unwin.

Scitovsky, T. (1954) 'Two Concepts of External Economies', *Journal of Political Economy* 62.

Scott, M.F. (1989) *A New View of Economic Growth*, Oxford: Clarendon Press.

SCRC (Social Science Research Center) (1987) *Pongjonghapjechul eui Gugmin-gyungje Giyu mit Giupmunhwa Yunku* (A Study of POSCO's Contribution to the Korean Economy and its Corporate Culture), Seoul: Seoul National University.

Sen, A. (1983) 'Development: Which Way Now?' *Economic Journal* 93: 745–62.

Shin, J.S. (1992) *Catching-Up and Technological Progress in Late-Industrialising Countries*, unpublished M.Phil. dissertation, University of Cambridge.

Smelser, N.J. (1976) *Comparative Methods in Social Sciences*, Englewood Cliff, N.J.: Prentice-Hall.

Smith, T.C. (1955) *Political Change and Industrial Development in Japan: Goverment Enterprise 1868–1880*, Stanford, Calif.: Stanford University Press.

Soete, L. (1985) 'International Diffusion of Technology, Industrial Development and Technological Leapfrogging', *World Development* 13/3: 409–22.

Solow, R.M. (1956) 'A Contribution to the Theory of Economic Growth', *Quarterly Journal of Economics* 70.

Solow, R.M. (1957) 'Technical Change and Aggregate Production Function', *Review of Economic Studies* 39.

Solow, R.M. (1959) 'Investment and Technical Progress', in Arrow, K. *et al.* (eds) (1959) *Mathematical Methods in the Social Sciences*, Stanford, Calif.: Stanford University Press.

Song, B.N. (1990) *The Rise of the Korean Economy*, Oxford: Oxford University Press.

Stiglitz, J. (1987) 'Learning to Learn, Localized Learning and Technological Progress', Dastupta, P. and Stoneman, P. (eds) (1987) *Economic Policy and Technological Performance*, Cambridge: Cambridge University Press.

Strachan, H.W. (1976) *Family and Other Business Groups in Economic Development: The Case of Nicaragua*, New York: Praeger.

Streeten, P. (1959) 'Unbalanced Growth', *Oxford Economic Papers* 11: 167–90.

Suh, Sang Chul (1975) 'Development of a New Industry through Exports: the Electronics Industry in Korea', in Hong, Wontack and Krueger, A.O. (eds) (1975) *Trade and Development in Korea*, Seoul: Korea Development Institute.

Supple, B. (1976) 'The State and the Industrial Revolution 1700–1914', in Cipolla, C.M. (ed.) (1973) *The Fontana Economic History of Europe: The Industrial Revolution*, Vol. 3, London and New York: Harvester Press.

Suzuki, Y. (1991) 'The Rise and Decline of Steel Industries: A Business Historical Introduction', in Abe, E. and Suzuki, Y. (1991) *Changing Patterns of International Rivalry: Some Lessons from the Steel Industry*, Tokyo: University of Tokyo Press.

Sylla, R. (1991) 'The Role of Banks', in Sylla, R. and Tonilo, G. (eds) (1991) *Patterns of European Industrialization: The Nineteenth Century*, London: Routledge.

Tiffany, P.A. (1991) 'The American Steel Industry in the Post-war Era: Dominance and Decline', in Abe, E. and Suzuki, Y. (1991) *Changing Patterns of International Rivalry: Some Lessons from the Steel Industry*, Tokyo: University of Tokyo Press.

Tilly, R. (1967) 'Germany 1815–70', in Cameron, R.E. (1967) *Banking in Early Stages of Industrialization*, New York.

Tilly, R. (1991) 'Germany', in Sylla, R. and Tonilo G. (eds) (1991) *Patterns of European Industrialization: The Nineteenth Century*, London: Routledge.

Tipton, F.B. (1976) *Regional Variations in the Economic Development of Germany during the Nineteenth Century*, Middleton, Conn.

Trebilcock, C. (1981) *The Industrialization of the Continental Powers 1780–1914*, London: Macmillan.

Ueno, H. (1980) 'The Conception and Evaluation of Japanese Industrial Policy', in Sato, K. (ed.) (1980) *Industry and Business in Japan*, White Plains, N.Y.: Sharpe.

Uenohara, M. *et al.* (1984) 'Background', in Okimoto, D. *et al.* (eds) (1984) *Competitive Edge: The Semiconductor Industry in the US and Japan*, Stanford, Calif.: Stanford University Press.

Vernon, R. (1966) 'International Investment and International Trade in the Product Cycle', *Quarterly Journal of Economics* 80/2: 190–207.

Verspagen, B. (1993) 'Convergence in the Global Economy: A Broad Historical Viewpoint', paper presented in the International Colloquium on 'Economic Growth and Structural Change: Comparative Approaches over the Long Run on the Basis of Reconstructed National Accounts', Leuven, Belgium.

Wade, R. (1990) *Governing the Market: Economic Theory and the Role of Government in East Asian Industrialization*, Princeton, N.J.: Princeton University Press.

Wallis, J.J. and North, D. (1986) 'Measuring the Transaction Sector in the American Economy, 1870–1970', in Engerman, S.L. and Gallman, R.E. (eds) (1986) *Long-Term Factors in American Economic Growth*, Chicago, Ill.: The University of Chicago Press.

Webb, (1980) 'Tariff, Cartels, Technology and Growth in the German Steel Industry', *Journal of Economic History* 40/2: 309–29.

Weber, M. (1949) *The Methodology of the Social Science*, translated and edited by Shils, E.A. and Finch, H.A., Glencoe, Ill.: Free Press.

Weber, M. (1961) *General Economic History*, translated by Knight, F.H., London: George Allen & Unwin.

Weinstein, F.B. *et al.* (1984) 'Technological Resources', in Okimoto, D. *et al.* (eds) (1984) *Competitive Edge: The Semiconductor Industry in the USA and Japan*, Stanford, Calif.: Stanford University Press.

Westphal, L., Kim, L. and Dahlman, C. (1985) 'Reflections on the Republic of Korea's Acquisition of Technological Capability', in Rosenberg *et al.* (eds) (1985) *International Technology Transfer*, New York: Praeger.

White, L.J. (1978) 'The Evidence on Appropriate Factor Proportions for Manufacturing in LDCs: A Survey', *Economic Development and Cultural Change* 27.

Whitley, R. (1992) *Business Systems in East Asia: Firms, Markets and Societies*, London: Sage.

Williamson, O.E. (1975) *Markets and Hierarchies: Analysis and Antitrust Implications*, New York: Free Press.

Williamson, O.E. (1985) *The Economic Institutions of Capitalism*, New York: Free Press.

World Bank (1990; 1991), *World Development Report*, London: Oxford University Press.

World Bank (1993) *The East Asian Miracle: Economic Growth and Public Policy*, London: Oxford University Press.

Yonekura, S. (1991) 'The Postwar Japanese Iron and Steel Industry: Continuity and Discontinuity', in Abe, E. and Suzuki, Y. (eds) (1991) *Changing Patterns of International Rivalry: Some Lessons from the Steel Industry*, Tokyo: University of Tokyo Press.

Yoo, J.H. (1990) 'The Industrial Policy of the 1970s and the Evolution of the Manufacturing Sector in Korea', KDI Working Paper no. 9017, Seoul: KDI.

Yoon, Jung Rho (1990) 'Hankook eui Sanup Baljun gwa Gugka: Bandoche Sanup eul Joongsim euro' (Industrial Development of Korea and the State: With Special Reference to the Semiconductor Industry), in Hankook Sahwesa Yunguhwe (ed.) *Hyundai Hankook ui Saangsanryuk gwa Gwahak Gisul* (Productive Capacity, Science and Technology of Contemporary Korea), Seoul: Moonhak gwa Jisungsa.

Yoshimori, M. (1991) Comment on Juhn, S. (1991) in Abe, E. and Suzuki, Y. (1991) *Changing Patterns of International Rivalry: Some Lessons from the Steel Industry*, Tokyo: University of Tokyo Press.

Zysman, J. (1983) *Governments, Markets, and Growth: Financial Systems and the Politics of Industrial Change*, Oxford: Martin Robertson.

# INDEX